THE FUTURE OF LESS

WHAT THE WIRE**LESS**, PAPER**LESS**, AND CASH**LESS** REVOLUTIONS MEAN TO YOU

ALLEN H. **KUPETZ**

EMERALD BOOK CO.

Published by Emerald Book Company
4425 South Mo Pac Expwy., Suite 600
Austin, TX 78735

For ordering information or special discounts for bulk purchases, please contact Emerald Book Company at: 4425 South Mo Pac Expwy., Suite 600, Austin, TX 78735, (512) 891-6100.

Design and composition by Greenleaf Book Group LLC

Publisher's Cataloging in Publication Data
(Prepared by The Donohue Group, Inc.)

Kupetz, Allen H.
   The future of less : what the wireless, paperless, and cashless revolutions mean to you / Allen H. Kupetz. -- 1st ed.

   p. : ill., charts ; cm.

   Includes bibliographical references.
   ISBN: 978-1-934572-09-2

1. Technological innovations--Management. 2. Business enterprises--Technological innovations. 3. Telecommunication--Technological innovations. 4. Finance--Technological innovations. I. Title.

HD45 .K87 2008
658.5/14   2008931898

Printed in the United States of America on acid-free paper

08 09 10 11 12 13   10 9 8 7 6 5 4 3 2 1

First Edition

CONTENTS

FIGURES

ACKNOWLEDGMENTS

I had been thinking about many of the ideas in this book for more than six years before I had the time and energy to convert those ideas into words on a page. In the interim, the realm of the possible has changed so dramatically that what I thought back then might someday come to pass has in many cases already become commonplace. I hope six years from now I'll look back on the ideas in this book and see that many of them are commonplace too.

Six years ago I started writing about what was then called *fourth-generation wireless*, long before third-generation wireless was available in the United States. My first thoughts on the topic were written on a x486 desktop PC running Windows 98. Now I can buy a mobile phone with more processing power than that PC. And a two-hundred-dollar iPod has eight times the memory. Certainly one of the greatest challenges in writing this book has been the accelerating pace of technological change.

At the risk of omitting someone, I must mention some of the many people who helped along the way. While no one named below can in any way be held responsible for the contents of this book, their contributions require grateful acknowledgment.

Since 2001, when I won a consulting contract with MeshNetworks, the CEO and CTO at that time, Richard Licursi and Peter Stanforth, respectively, helped shape my vision of a wireless future. Richard created an environment where visionaries could turn ideas into products. And as I travel around the country talking about *The Future of Less*, I often quote the Stanforth Rule: "There are only two technological problems that exist in the world—those that violate the laws of physics and those that time and money can solve." K. Terrell (Terry) Brown, a marketing manager at MeshNetworks, worked with me to put some of my early ideas on paper, some of which were used in this book. Martin Suter, then MeshNetworks vice president for strategic business development, was always a useful sounding board and reality checker and also helped me straighten out my golf drives. Fernando Arozqueta has been my friend and teacher for more than ten years and is an accomplished sales professional. Like Martin, he will deny ever having tried to help my golf game.

Several colleagues at the Crummer Graduate School of Business at Rollins College also provided encouragement and support. Professors Ilan Alon, Susan Bach, Marc Fetscherin, James Gilbert, James Higgins, James P. Johnson, Mark Johnston, Tom Kruczek, Tom Lairson, Marc Sardy, and Clay Singleton must be singled out in writing. My colleague and weekly poker opponent, Dr. Barry Render (a prolific and successful author in his own right), red-penned an early draft of this and really helped me to sharpen my focus. The dean of the Crummer School, Dr. Craig McAllaster, and the entire faculty approved the creation of a new course I wanted to teach—Managing Technology—and the interaction that allowed with the kinds of students that Crummer attracts gave me several new ideas. Dean McAllaster also was very helpful in getting this book published. Serving as Crummer's executive-in-residence since 2005 is probably the best job I've ever had.

These folks have—at the very least, inadvertently—helped shape my ideas: Steve Alkhoja, Chaim Amram, Kimon Anemogiannis, Gene Apelado, T. Lamine Ba, Mark Bartschi, Roe Bibona, Doug Campbell, Eric Choi, Hasok Chun, Dr. Lewis Duncan, Frost & Sullivan Asia Pacific (Manoj Menon, Koo Hee-sang, and Jafiz Ishahak), Paul Gilgallon, Glen Gray, Josh Hallett, Patty Hamm, Dan Hannah, Marc Hayden, Phil Hill, Mack Hooper, Yo Hoshino, Mike Johnson, Iain Juden, Joel Karp, Larry Koos, Paul Latchford, Joseph Lee, Michal Lev-Ram, GC Liang, Thomas Liu, Eric Love, Jim McSorley,

Dr. Joel Melnick, Dennis Pape, Jody Pendleton, Bai Ray, Allen Roberson, Robin Roberts, Scott Rogers, Trey Roper, Rick Rotondo, Michael Rubin, Susan Storma, Fernando Verrua, Linda Wang, Shawn Welsh, Stephen Wildstrom, James Wong, and Tom Yang. All these people understand a lot about technology and, more importantly, the difference between an idea, a technology, and a product—something I try to teach my students. And I hope by including all their names that they will buy multiple copies of this book. They have all earned a free copy to be sure, but I think they will appreciate this book all the more if I make them buy their own. A special thanks to Eric Love for his graphics and to Alex Bourque for the creation of www.futureofless.com.

My fifteen-year-old son, Allen J. Kupetz, read a draft of this cover-to-cover and found—much to his delight—an embarrassingly large number of typos. My twelve-year-old daughter, Leslie, also must be thanked for occasionally dragging me away from this text on my laptop for a game of Ping-Pong or a swim in the pool.

I first visited South Korea in 1992, and I remain impressed that such a small country has accomplished so much. Spend a week in Seoul and you'll know that no place in the world has such a great concentration of talented human capital. Special thanks to Chris Ahn, Choi Jin-Young, Chung Ho-Young, Dr. Jeon Jung-Ok, Kim Byung-Jung (B.J.), Kim Dae-Hee, Dr. Kim Gihong, Dr. Kim Hansuk, Kim Hee-Tae, Kim Yoon-Chung, Richard Kwon, Lim Seung-Chan, Park Jae-Surn, Park Hyeon-Jin, Dr. Seo Jung-Uck, Seo Sangwon, Woo Tae-Hee, and Yeo Han-Koo.

Even with all the input from the thinkers and writers above, the text that follows would be vastly inferior without the help of my sister, Marilyn. Her work extended way beyond commas and capitalization of the earliest pieces of the first draft; she helped me write in such a way that I hope makes this easy for readers to enjoy. That said, I don't always listen, so any mistakes in this book are mine and mine alone.

Great thanks to Clint, Meg, Lari, Alan, Jay, Erin, Lisa, and the entire Greenleaf Book Group team. What a pleasure it was to work with all of you. And I didn't always listen to you all either, so you are also not responsible for my mistakes.

Finally, nothing I do would have been possible without the simple plan my parents, Jerry and Joan, had for me: get a good education, treat others honestly and fairly, and trust that most everything else will take care of itself. They funded my education, giving me a chance to focus on school. Both life-long learners and voracious readers, they helped make sure my future would not be one of less.

PREFACE

You say you want a revolution
Well you know
We all want to change the world . . .
Don't you know it's gonna be alright.
—JOHN LENNON AND PAUL MCCARTNEY

The Future of Less is about revolutions: the wireless, paperless, and cashless revolutions that are already well underway, hugely beneficial, and arguably unstoppable as the technologies behind these revolutions make our lives more convenient, more in tune with the environment, and safer. This book is also about the kinds of ideas that lead to products that people are buying today—and using to change virtually every aspect of how they communicate with each other. What follows are four collections of thoughts bound into a single book. I hope the pages on the wireless, paperless, and cashless revolutions will get you thinking, among other things, about how you can use these revolutions to simplify your life and make technology serve you rather than you serving it. The fourth piece is a look at these revolutions around the world, focusing on some of the most advanced nations to implement them (Japan and South Korea), the up-and-comers (China, India, and the Philippines), and those handful of countries in Africa that can only now see the technological revolution on the horizon and are already preparing for it.

My perspective on these revolutions has been influenced perhaps most strongly by my experience in South Korea. I lived in Seoul from 1992 to 1996 while serving as the telecommunications policy officer at the U.S. embassy. This period included the launch of the world's first commercial next-generation wireless network, which helped lay the foundation for Korea to become one of the most advanced countries in the world in terms of IT deployment. South Koreans are early adopters of new technologies and quickly weed out those that provide little value. After this informal testing period, the surviving technology goes global and we get to use it here in the United States. I'm looking forward to my refrigerator telling me the milk is past its expiration date and using my cell phone to pay for public transportation like you can in Korea today. Therefore, I chose to use what is going on in South Korea as a reference throughout the book to give readers a look into the future. I continue to closely follow political, economic, and technical developments in Korea and my blog can be found at www.koreality.com. I hope you find the glossary useful. Despite my best efforts, there is some technical and business jargon in the pages that follow. I also tried to use terms consistently. For example, most of us use the term *cellular company* to describe the organization we send our monthly payment to in exchange for mobile voice service. But as we start to move away from solely cellular networks, and continue to buy phones that do a lot more than just allow us to talk, we need something more accurate. Newspapers and magazines sometimes use *cellular operator*, *cellular carrier*, *mobile service provider*, *mobile operator*, or *wireless carrier*, just to name a few. I chose the term *mobile network operator* (MNO), which I think encompasses all of the above. Likewise, I've tried to avoid using the term *cell phone* and instead have mostly used *mobile phone* or *mobile device*.

I want you to enjoy this book. I hope it helps you think about what wireless, paperless, and cashless mean to you and will mean to you in the years ahead. And I'd like you to challenge any of my ideas via my blog at www.futureofless.com.

But I also want you to take these ideas and use them to change the way your business is doing business. The wireless, paperless, and cashless revolutions are already impacting both sides of your supply chain (vendors and customers), your employees, your marketing plan, and your physical office. What is your business doing to grow and adapt and to get ready for the future?

The two ends of the emotional spectrum of the impact of wireless, paperless, and cashless revolutions are excited/optimistic and nervous/confused. Most of us fall somewhere in between. However you feel about the future of less, I hope we can agree on these basic points: the wireless, paperless, and cashless revolutions are already well underway, unstoppable, and will have profound impacts on virtually every aspect of how we work, live, and play.

INTRODUCTION

We live in a moment of history where change is so speeded up that
we begin to see the present only when it is already disappearing.
—R. D. LAING

Where are the flying cars?
For me, born in the early 1960s and one of the millions
who watched all the Apollo moon landings, the prediction of the
future that stands out in my mind best is that we would all be commuting in
cars that flew. No more traffic jams. No more toll roads. Perhaps they all had
small nuclear reactors in them so there was no need for gas or any other fuel.
The plan was for *The Jetsons* to look like a reality television show.

Almost forty years after man first landed on the moon, my car is still
permanently land-based, I still pay tolls, and I'm still filling it with gas.
What did the promised transportation revolution deliver instead? The Segway. The company first marketed itself as "the next generation in personal
mobility,"[1] which apparently meant traveling at about 12.5 miles (20 km)
per hour with a range of 24 miles (38 km) before you had to recharge the

---

1   See http://www.segway.com/products/.

lithium-ion battery packs. As revolutionary as *The Jetsons*? This can't even compete with the Batmobile.

The inventor of the Segway, Dean Kamen, once predicted that "the Segway will be to the car what the car was to the horse and buggy."[2] I remember the buzz that the Segway would mean the end to walking. It is hard to support the idea that the next transportation revolution should be the end of walking.

Dean Kamen is a smart guy—certainly smarter than I am. But the Segway was obviously designed in a vacuum, devoid of consumer input. The "if you build it, they will come" model of product development—focused on engineering rather than consumer research—is based on hope and, as the cliché goes, hope is not a strategy. So I try to contrast my ideas—whether you think they are any good or not—with those of engineers, like Kamen. Since I'm not an engineer, I consider myself instead a *technologist*, which I define as such: the optimist sees the glass half full; the pessimist sees the glass half empty; and the technologist wonders why the engineer didn't talk to anyone before building a glass that was twice as large as anyone wanted or needed.

THE FUTURE IS HERE

Much of the rest of the world is way ahead of the United States in embracing and implementing the wireless, paperless, and cashless revolutions I describe in this book. As the science-fiction author William Gibson is credited with saying, "The future is here. It's just not widely distributed yet." The reasons for this are often cultural or regulatory, not technological. For the United States, the good news about being the second or third mover in the marketplace is that it could offer us a chance to catch up eventually by eliminating the need to try technologies that failed elsewhere. And by the time we implement these technologies, some economies of scale may have kicked in and dramatically lowered their costs, both for MNOs and for consumers.

Culture and regulatory issues change more slowly, but change does come. The Internet and the World Wide Web are case studies in how technology can change culture and force governments to embrace a new regulatory vision.

2   Rivlin, Gary. (2003, March). Segway's Breakdown. *Transportation Alternatives*. Retrieved May 12, 2008, from http://www.transalt.org/press/media/2003/030301wired.html.

Can you imagine the government not trying to tax anything and everything it can? Yet most interstate e-commerce transactions in the United States are not taxed. It is remarkable. And not too long ago filing your income taxes online was impossible from a regulatory standpoint, even though the technology had been around for more than a decade. Regulatory change does come, however slowly, when people demand that the government allows them to do what technology allows them to do.

So it is with good ideas, technology, products, and services—they can change as people change and as technologies around them change. Of the three revolutions I address in this book, the most important one is the wireless revolution, because it enables and enhances the other two. Man is inherently a mobile creature, and wireless technology—coupled with the Internet and the World Wide Web—allows us to store and retrieve information anywhere, anytime, on the move. The wireless revolution enables a modern world where BlackBerrys are more commonplace on a belt than in a cobbler. But certainly wireless isn't important just because it is making the migration to a more paperless and cashless society possible. The mobile phone is the most sought-after consumer electronic product globally, ahead of the personal computer or the car. In many parts of the world, a mobile phone is the user's personal computer. I have been to more than thirty countries on six continents, and only the near ubiquity of the mobile phone is common everywhere I go.

The second revolution can seem less welcome. For many of us, going paperless seems like something we should do or something we have to do, not something we want to do. It's like eating vegetables as a child, and most of us never volunteered for lima beans. For the first of what will be many times in the pages that follow, let me state the following clearly: paperless doesn't mean no paper. The overwhelming majority of you reading this are doing so with a device that doesn't need electricity to run, starts instantaneously, and never needs to be rebooted: a book. The irony of writing a book about going paperless is not lost on me.

Nonetheless, the paperless revolution is important for two reasons. Paperless communication allows for the digitization of ideas, which makes them faster and easier to share across distances. If the global citizenry is capable of accessing information using mobile phones, then let's not only print information in big, heavy, expensive books. Let's also put it on the World Wide Web so the broadest possible audience has access to that information at the lowest possible price. The paperless revolution is also inherently eco-friendly, since it means cutting down fewer trees, using fewer chemicals and less energy in recycling paper, and fewer printer ink cartridges to dispose of.

While writing this book, I tried to complete the paperless section paperlessly. Instead of using the usual research techniques of getting books, photocopying pages, and clipping magazine articles, I did almost everything digitally. I stored the research on my computer (I converted the articles to .pdf just in case the content disappeared from the Web) and did my edits onscreen rather than with a red pen on printed paper. It was not easy at first, as old habits are hard to break. But as someone who travels a lot, I enjoyed the convenience of always having all my research with me without having to drag around manila folders stuffed full of articles and photocopies. But paperless is possible and, like eating most vegetables, you should probably do it whether you like it or not. I still wouldn't volunteer for lima beans.

Of the three revolutions under discussion, the cashless future might be the easiest to envision given the huge role that credit and debit cards already play in the U.S. and global economies. The notion of a cashless society is appealing because I think the end of cash will mean a huge reduction in cash-related crimes like muggings and bank robberies. No cash means no cash to steal, no money for the government to mint and for us to lose, and no more having too much or too little change. Cashless means convenience and safety.

The ideas, technology, products, and services that are part of the wireless, paperless, and cashless revolutions have changed and will continue to change. What we know today as cellular phone technology was designed solely to talk to other people, not text message, get driving directions, or access the World Wide Web. Life without paper hasn't been viable since shortly after Gutenberg commercialized movable type. And to almost everyone in the United States—at least those who haven't read this book—using a credit or debit card at the register is the full extent of the cashless revolution. But change is coming, and with a look to the early-adopter culture of South Korea (included in each chapter of the book) we can get a pretty good idea of what it will be. The question is now only this: are you ready for the revolutions?

SECTION 1

THE FUTURE OF WIRELESS

CHAPTER 1
FROM WIRED TO WIRELESS

No sensible decision can be made any longer without taking into account not only the world as it is, but the world as it will be. —ISAAC ASIMOV

I t took more than a century to populate the planet with 1 billion telephones. The second billion took only five years. By the end of this decade, 2 billion people will own not just phones, but phones with web browsers.[3]

When you left your home today, what did you make sure that you didn't leave home without? My guess is you carried money (wallet or purse), keys, and your mobile phone. Mobile voice and data communications technology—what we generally refer to today as cellular networks and phones—is proving to be the most important consumer product developed in my lifetime. It is even more important than the personal computer, because the mobile phone shows the promise of becoming the personal computer for billions of people who couldn't otherwise afford one. Wireless technology enables humankind's inherent desire for mobility. The global information revolution brought about by the Internet and the World Wide Web, coupled with mobile wireless technology, has changed the world forever.

---

3   Karlgaard, Rich. (2007). *Our Challenge Is Change, Not Globalization.*

Is it just hype? People seemed to exist just fine on the planet for a very long time without mobile phones. But like planes, trains, and automobiles, like sliced bread, 99-cent cheeseburgers, and Starbucks coffee, some products and services simply transform societies. Wireless technology has transformed the world. There are more people using mobile phones in China today than there are people in the United States. There are hundreds of millions of people in the world who have never made a traditional wired phone call and never will. Nor have they ever dialed up the Internet or connected to the World Wide Web with the wires of digital subscriber line (DSL) or cable service providers. These millions—arguably these billions—of people will never use anything other than wireless technology for all their phone calls, Internet browsing, and eventually television.

First-generation (1G) wireless telecommunications—the bricklike analog phones that are now collector's items and sold only on eBay—introduced the cellular architecture in the United States in 1982 that is still being offered by most wireless companies today. While the system worked well, it allowed only one user per radio channel, similar to the way only one person can use the wired line in your home at a time. Subscriber demand quickly overwhelmed the capacity of the technology. Second-generation (2G) wireless, launched in 1990, supported more users within a cell, the area surrounding the cell tower that a cell phone is connected to wirelessly. 2G wireless used digital technology, which allowed three to eight users to use the same channel. But 2G was still meant primarily for voice communications, not data, except some very low data-rate features, like short messaging service (SMS, also known as *text messaging*). In the late 1990s, so-called 2.5G allowed cell phone companies (referred to hereafter as *Mobile Network Operators*, or MNOs) to increase data rates with a software upgrade at their cell towers, as long as consumers purchased new phones. These data rates offered service at up to 384 kilobits per second (Kbps), about seven times faster than the average dial-up modem.

Early in this century, third-generation (3G) wireless was introduced in Japan and South Korea and offered greater bandwidth, basically bigger data pipes to users, allowing them to send and receive more information. The United States didn't see true nationwide 3G coverage until 2007. Japan and South Korea, among the first countries with nationwide 3G coverage, are not surprisingly leading the world with so-called 3.5G networks based on high-speed downlink packet access, better known by the marketing-challenged acronym HSDPA. HSDPA is essentially a way to increase the data rates of early 3G networks—a subscriber can download more information in a shorter period of time. In Japan and South Korea, subscribers with HSDPA are reportedly seeing data rates in excess of DSL rates in the United States. Others in Asia are well down the path too. Europe is trailing most of Asia, but is ahead of the United States. Even in Latin America, some countries are ahead of the United States in terms of wireless broadband.

GETTING UNPLUGGED

Wire telegraph is a kind of a very, very long cat. You pull his tail in New York and his head is meowing in Los Angeles. And radio operates exactly the same way: you send signals here, they receive them there. The only difference is that there is no cat. —**ALBERT EINSTEIN**

Nicholas Negroponte, the founder of MIT's famed Media Lab, is credited with first asking in the 1990s why we used to deliver low-bandwidth applications (the sending/receiving of small amounts of data, like voice) over high-bandwidth wired networks, while we were delivering high-bandwidth applications (like television and other video) over low-bandwidth wireless networks like UHF and VHF (which used to be delivered via a big antenna on your roof before cable television). The answer is that the early pioneers of both of these applications—voice and video—never imagined that computers would make mobile wireless voice possible and inexpensive, or that users might ever demand more than three or four television channels.

> **MIGRATION** The migration from first-generation wireless networks to where we are today has been slow, steady, and entirely predictable. As we wanted to use the same bandwidth-intensive applications that we could from the home or office over wires, wireless networks needed to evolve to give us more bandwidth. Better and better mobile bandwidth plus mobility has resulted in the rapid growth we've seen in mobile

> service around the world. For your planning purposes, it is safe to assume that everything you can do while you are wired to the Internet you will be able to do with a mobile phone using wireless.

This Negroponte Shift describes the migration from devices that traditionally used wires (e.g., telephones) to wireless (e.g., mobile phones) and the roughly simultaneous migration of devices that traditionally were wireless (e.g., television) to wire (e.g., cable television). Negroponte correctly predicted that wireless networks would soon become robust enough to handle voice and some amount of data. And that television, more often than not viewed by someone sitting down, didn't need the mobility that comes from wireless, but did need the added bandwidth available from wire-based delivery systems.

People are mobile creatures. They often need to be mobile (to go to/from work) or want to be mobile (to go to the grocery store with their mobile phone when expecting a call from their child, instead of waiting at home). So people use wireless to allow them to be mobile while still doing many of the things they formerly could do only at work or at home. The advancement of technology has meant ever-increasing data rates for wireless networks. In Asia, lots of video content, including television, is already delivered over cellular and other mobile networks to mobile phones. Negroponte wasn't wrong in the 1990s; he just didn't foresee that virtually all content eventually would be—or at least would be able to be—delivered wirelessly. Ubiquitous networks—those able to deliver any content to anyone anytime and virtually anywhere—are already becoming a reality in urban centers in Asia and will eventually be truly ubiquitous. We used to talk about time shifting—the power that VCRs and now digital video recorders like TiVo gave us to record our favorite programs and watch them *when* we want to. Now we are seeing place shifting—the ability to watch content *where* we want to.

Figures 1.1–1.3 clearly show the global number of traditional wired subscribers flattening or decreasing. Even in developed countries, there is no

longer a need for traditional wired phone service. It is likely that the majority of so-called *millennials*—Americans born after 1982 who graduated high
school in the new millennium—will never have traditional phone service or
ever use a pay phone.

Figure 1.1 Fixed vs. Mobile Penetration Rates—Americas (1996–2006)[4]

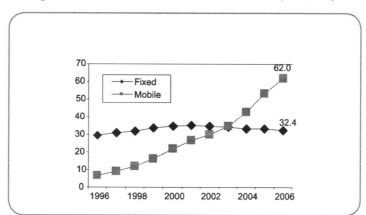

Figure 1.2 Fixed vs. Mobile Penetration Rates—Asia-Pacific (1996–2006)[5]

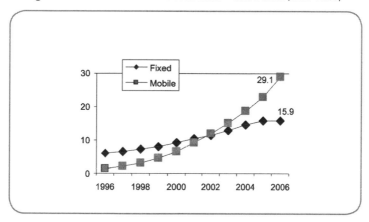

4   ITU. (2006). Retrieved May 12, 2008, from http://www.itu.int/ITU-D/ict/statistics/ict/graphs/am1.jpg.
    Reproduced with the kind permission of the ITU.
5   ITU. (2006). Retrieved May 12, 2008, from http://www.itu.int/ITU-D/ict/statistics/ict/graphs/ap1.jpg.
    Reproduced with the kind permission of the ITU.

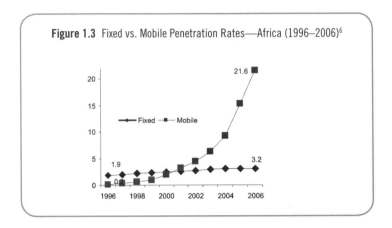

Figure 1.3 Fixed vs. Mobile Penetration Rates—Africa (1996–2006)[6]

Video on demand (VOD) will be the next step in the wireless revolution. There was a television commercial from a regional U.S. phone company in 2000 promoting its yet-to-be-released VOD service that had a man walking into a run-down motel:

"Do you have king-size beds?" he asked.

"No," was the reply from the unfriendly-looking and-sounding clerk.

"Breakfast?"

Again, "No."

"Well, what do you have?" he asked in frustration.

"We have every movie ever made in any language, all day long," the clerk responded.

"How is this possible?" the customer asked in disbelief.

Traditional phone and cable companies have been promising for years to roll out VOD. The service would allow you to view a list of titles, select the one you want, and play it right then. Eight years after that commercial, VOD still hasn't been fully implemented by the phone and cable companies, but most offer some sort of service with a limited number of titles. Netflix, the DVD-by-mail pioneer, also has a limited VOD service delivered over the Internet.

Let's return to the idea that people are mobile creatures. Most people at home would prefer to watch movies on the biggest screen in the house, which is not their mobile phone. But for those Japanese using public transportation

---

6   ITU. (2006). Retrieved May 12, 2008, from http://www.itu.int/ITU-D/ict/statistics/ict/graphs/af1.jpg. Reproduced with the kind permission of the ITU.

for their notoriously long round-trip commutes every day, VOD to mobile phones is an obvious business and why that country is leading the world in adopting this service.

THE PATH TOWARD UBIQUITOUS NETWORKS

Technology is stuff that doesn't work yet. —**BRAN FERREN**

In March 2005, I took six of my MBA students to South Korea to study the wireless revolution (I've taken more than a hundred since then). Before we left I told them that Korea was three years ahead of the United States in terms of wireless networks. I was wrong. After three years, the wireless services and features I saw there then are *still* only dreams in the United States. It's now clear to me that South Korea was then—and is now—almost five years ahead of the United States. A wondrous combination of savvy, demanding customers; MNOs that aggressively compete for those customers; and intelligent regulatory oversight has made Korea a leader in the wireless revolution.

During visits to the Ministry of Information and Communication, MNO offices, and mobile-device makers in and around Seoul in 2005, one word that kept appearing (and is still a dominant marketing theme today) was *ubiquitous*. But what does ubiquitous mean? In Korea, ubiquitous means that a subscriber should be able to access the Internet anywhere at any time— anything you could do at the home or office while connected via a wired broadband connection you should be able to do on the subway with a laptop computer or handheld device. And your user experience should be identical in terms of applications, data rates, and reliability of the connection.

The goal of the government in promoting the creation of ubiquitous networks was to increase productivity—people would be able to work during their commutes. The goal of the MNOs was more billable minutes—people would use their mobile phones more and would pay for new applications. And the goal of the device makers was to create demand for new and innovative mobile devices. Given enough time and money, to paraphrase the Stanforth Rule, something this possible and useful to deploy will eventually be deployed. Something approaching ubiquitous access will almost certainly be available in every developed country by 2020, probably first in the geographically smaller countries. Japan, Singapore, and South Korea are approaching ubiquitous networks today.

South Korea's wireless leadership isn't surprising, however, when you consider the connectivity already at play in South Korea. Let's take a look at the numbers.[7] In terms of broadband, both wired and wireless, 90 percent of South Korean homes have broadband Internet access. The world average is about 20 percent. 100 percent of South Korea's wired Internet access is broadband (i.e., dial-up has disappeared). The world average is about 30 percent. Over 50 percent of South Koreans have migrated their mobile phone to 3G service. The world average is 5 percent. Korea has over 25,000 commercial Wi-Fi hot spots. The highest broadband data rates in South Korea in 2006 were already more than three times faster than in Canada.

- As words of J. Bradford DeLong wrote, "For the past century, people around the world have looked to the United States to see what their own futures will be like . . . except where broadband is concerned. In this respect, we need to look at South Korea."[8] Here's what Koreans are doing with this broadband access:

- 6 percent of South Koreans make payments using their mobile phones. The world average is under 5 percent.

- 57 percent of South Korean music sales were digital in 2006, versus 10 percent in the United States. 26 percent of South Koreans listen to music on their mobile phones, versus 4 percent in the United States.

- 37 percent of South Koreans download games to their mobile devices, versus 10 percent in the United Kingdom. 15 percent of South Koreans play video games on their mobile phones every day.

- 20 percent of South Korean mobile subscribers use an Internet search engine on their mobile phones. 14 percent check the weather that way.

- 40 percent of South Korean youth text message in class, with 33 percent of them sending over 100 text messages per day.

- 30 percent of South Koreans upload pictures from their camera phones to social networking sites, versus 10 percent in the United Kingdom.

---

7   Ahonen, Tomi, and O'Reilly, Jim. (2007). *Digital Korea.*
8   DeLong, J. Bradford. (2003). Seoul of a New Machine. *Wired*, p. 83.

- 42 percent of South Koreans use MMS (i.e., like text messaging, but with pictures or video), versus 19 percent of Germans. And 97 percent of South Koreans buy ring tones, versus 7 percent of Germans.

- 43 percent of South Koreans maintain a blog or social networking profile, versus 21 percent in the United States. The industrialized world average is about 10 percent.

That South Korea is the first or among the first countries to achieve such a high level of broadband access—wire and wireless—could easily be dismissed as simply a logical consequence of a population living in dense urban environments in a geographically small country, both of which greatly reduce the cost of deploying broadband networks. Certainly Korea is small and highly urbanized, and certainly these factors have contributed significantly to its path toward ubiquitous networks and the wireless, paperless, and cashless revolutions such networks enable and foster. Mix in the cultural factors of an emphasis on education, an affinity for gadgets, and high disposable income, and South Korea's success seems likely. When you also factor in government policies that promoted growth across virtually every wireless vertical market, Korea's global leadership position was all but guaranteed.

South Korea used its geographic, cultural, and regulatory environment to roll out wireless applications in the early 2000s (e.g., television, GPS, and ring tones on mobile phones) that came to the United States only recently. Thus it is probably safe to conclude that what we are seeing in South Korea today (wireless VOD, mobile banking, and e-books) will be coming to mobile phones in the United States in the years ahead.

Does what is happening over there affect what will happen here? Isn't the United States the technology innovator, with Asia just focused on lowering the cost of producing our innovations? From cars to consumer electronics, this is how many Americans generalize the innovative contrast between Asia and the West.

But wireless technology is different. Asians, from both the biggest and wealthiest countries to those still with emerging economies, are the innovators as well as the low-cost manufacturers. Certainly Europe has its innovators too, and Nokia of Finland remains the world's dominant handset manufacturer, with almost 50 percent of global market share. But as handsets become more and more commoditized, it is in applications for mobile users that future innovation will be rewarded most richly. And the most innovative applications are coming from Asia.

There was a time not long ago that MNOs in the United States largely dismissed Japanese advances in the industry. Asians might want to send text messages, watch video on their phones, or buy ring tones, but not Americans, some MNOs believed. As it turns out, however, we like to do all those things, too. Companies (and not just Asian companies) that bet on the cross-cultural appeal of communications applications thrived. In the United States, Nokia's handset market share and Deutsche Telecom's T-Mobile successful hot spot business are testimonies to this. In South America, most of the largest MNOs are subsidiaries of Spanish, Mexican, and Italian companies. And Britain's Vodafone is now a global MNO. So, what happens over there is affecting—and will continue to affect—what is happening here.

UBIQUITOUS CONNECTIVITY for handheld mobile devices will mean that we will be able to access everything on the World Wide Web anywhere, anytime we want. This will have a huge impact for companies. Will employees be expected to be available 24–7? If not expected, will they be pressured to be available anywhere, anytime? What about your customers? What will their expectations be? Successful organizations will need new processes to deal with all their always-on stakeholders.

CONCLUSION

We should all be interested in the future because we have to spend the rest of our lives there. —**CHARLES F. KETTERING**

To those who don't know any better, the wireless revolution looks alive and well in the United States. Handheld mobile devices are relatively inexpensive. We have a choice of several different MNOs. We can run into a 7-Eleven and buy a prepaid phone for about $30. We can text message. Voice quality is pretty good, even though the guy sitting next to us in the restaurant feels he must shout into his phone to be heard.

Talking is what most Americans do with their phones. And why not? *Phone* comes from the Greek *phonos*, meaning "sound." But when I say we are still waiting for wireless, I'm talking about the complete migration from the idea of a cellular phone to a mobile device. A device that will be the Swiss Army knife of our life, able to perform dozens of functions—including making voice calls. That mobile device will connect us to the Internet anywhere and at any time. It will start our car and unlock the front door of our home. It will replace our wallet, MP3 player, camera, video camera, wristwatch, and GPS receiver. We'll watch television on it when we can't find a larger screen. It will be the computer of choice in the developing world. It will be how we communicate with everyone, whether or not we choose to talk to them.

TRENDS Trends in consumer electronics—cost, price, computing power—are also impacting mobile networks. Lower costs are achieved as MNOs reach economies of scale faster. They can pass that savings to consumers or use the extra profits to roll out new wireless products and services faster. The continued geometric growth in computing power that we see in personal computers will mean more powerful handheld mobile devices. If the past to the present was predictable, the future is too. The technology and cultural change we see in countries with the most advanced wireless networks—Japan, Singapore, and South Korea—will be available in the United States in three to five years.

For people choosing stocks to invest in through charts of their price and volume, there is a cliché: "The trend is your friend." Warren Buffett has made billions proving this is not always true when the trend is downward; and the burst of the tech stock bubble in 2000–2001 proved that upward-sloping trends are not sustainable forever either. But in technology, trends seem far more stable as we continually see items getting faster and cheaper in fairly predictable ways. So too with wireless trends.

Efficiency in wireless spectrum basically means getting more data packed into the same amount of bandwidth. Think of this as changing your brand of gasoline and getting double the miles per gallon. So certainly this leads to lower delivery costs. MNO rates started at about $5/minute and now, only about twenty years later, long distance is free to most users. The cost savings MNOs have seen from greater efficiency in their use of wireless spectrum have translated into cost saving for wireless subscribers. This is not the case of long-distance calls made over wires. And Moore's Law, which predicts a doubling in the number of transistors and therefore the processing power of computer chips every eighteen months, similarly forecasted the overall growth in information technology. An iPhone today has the processing power that a supercomputer the size of a room had just a decade ago. What these figures mean collectively is that we have moved from the digital revolution, which focused on digitizing documents, photographs, and records so that they could more easily be manipulated, to the wireless revolution: making digital information about anything available anywhere at almost no incremental cost over your monthly mobile-phone bill.

The migration from wires to wireless means—most importantly—mobility and ubiquitous access to everything the World Wide Web has to offer. Just as wired broadband to the home enabled workers to telecommute (less traffic) or simply work either more or at all hours of the day or night—or both (less work-life balance)—true mobility will have the same or more impact on our society and culture. It was not possible to see how the Internet would transform so much of our lives. But taking the same Internet and coupling it with the inherent mobility of people promises even more transformational change. The home office will become the anywhere-I-want-to-be office. We won't need to get to the bank before it closes because the bank will be in our pocket. And every book ever written will be instantly available literally at the touch of a button.

E-mail turned out to be the killer application for personal computers—the reason everyone had to have a PC. There has been a similar discussion about wireless: what is the killer application for handheld mobile devices? Certainly mobile voice was initially the driving force behind new subscriber growth, but now? The answer was too obvious for us to see it for many years. As it turns out, mobility is the killer application. It is not *individually* e-mail, or movies, or text messaging, or even voice calls that continue to drive the wireless revolution, but a combination of all of these in a mobile environment. Mobility is the cause of the wireless revolution. People want to be able to do what they want when and where they want.

Revolutions occur for a reason, and this holds true for the wireless revolution. E-mail was the cause for the personal computer revolution—consumers needed to buy a computer to send and receive e-mail.

Cars gave us mobility and caused a revolution. Airplanes increased mobility and caused a revolution. More people own mobile phones than own cars or personal computers. More people use them to connect to the world than fly on planes. In many developing countries, huge percentages of the population will never drive, fly, or own a traditional computer. Many countries in Asia now show over 100 percent wireless penetration and there are twice as many phone users as wired Internet users. Mobility, and the wireless revolution that changed the definition of what it means to be mobile, offers us a customization no other product does: the ability to use it when and where we want.

CHAPTER 2
PEERING INTO THE
WIRELESS FUTURE

When a distinguished and elderly scientist says something is possible, he is almost certainly correct; when he says something is impossible, he is very probably wrong. —**ARTHUR C. CLARKE**

W hat's wrong with the way we've always done it? As the number of wireless subscribers increases and the number of high-bandwidth applications increases, why not just build more and more cell towers to satisfy demand? To paraphrase Yogi Berra, "Sometimes it gets late early." We are already way past the number of subscribers the original designers of our current cellular architecture could even imagine. We need a new way to connect all these new users and applications—unless you want to volunteer your backyard and those of many of your neighbors so the MNOs can build new cell towers.

Given all the growth in the number of wireless subscribers, and the ever-rising expectations we have about where our mobile phones should work, it is not surprising that MNOs are driving around the country asking, "Can you hear me now?" If no one on the other end answers that question, it means the MNOs must put up another tower to offer service in that area. Wireless coverage depends on a signal from one of the roughly 200,000 towers in the United States today. But the cost of that tower is tremendous. MNOs must buy or rent the land, or rent the top of a building, install the tower or put their antenna on an existing tower owned or used by a competitor, and connect that tower back to their existing network. This can be an expensive and time-consuming process. And all those costs must be reflected on your cell-phone bill. A less costly way to do this would benefit both the MNOs and their customers.

In addition to lower mobile-phone bills, the differences between the kinds of network technology in use can have a considerable impact on the connectivity of the network's users. Less efficient networks are slower, less reliable, and less likely to support the mobile demands of their users. Have you seen those television commercials where one person is talking, says something he thinks is funny, only to get absolute silence on the other end? He fears he has offended the person he is talking to. Better networks would mean fewer awkward moments like this. Therefore, the kind of network infrastructure MNOs choose can affect not only their business, but also the lives of all those in their region. Ultimately, the speed with which the wireless revolution unfurls in the United States depends a good deal on the technological choices of the MNOs that serve it.

Our current mobile networks, whether they are 1G, 2G, 2.5G, or 3G networks, are cellular systems: a series of slightly overlapping circles or cells with a tower in the middle and boundaries based on the range of each tower. Mobile-phone users within a cell access the network via that tower. The circles overlap so that if you are driving, for example, you can connect to a new tower before you get out of range of the old tower. If you drive to where there is no tower in range, or where that tower has no additional capacity at that moment, you get a dropped call and must redial.

The advantage to cellular technology is that given enough time and money, MNOs can build nationwide networks, which most of the big MNOs have done. But there are disadvantages too. The tower in the middle is a single point of failure, meaning that if that tower is knocked over by a hurricane, for example, none of the people in that cell will be able to use their phones. A second problem is that there is little load balancing in cellular networks; a busy tower cannot ask another tower to share its workload. The single biggest disadvantage to cellular networks going forward is that more and more people are sending more and more data via their phones, and the cellular towers each have capacity limits. As data rates increase, the size of the cells (i.e., the diameter of those overlapping circles) will have to decrease to support those higher data rates. Smaller cells, however, mean more cell towers and the not-in-my-backyard battles that such proposals engender because of the fear that wireless towers cause cancer, interfere with other wireless devices, or are just ugly to look at.

WHAT SHAPE WILL YOUR NETWORK BE?

Any sufficiently advanced technology is indistinguishable from magic.

—ARTHUR C. CLARKE

One vision of what might become the fourth-generation (4G) wireless network was originally conceived by the Defense Advanced Research Projects Agency (DARPA), the same organization that developed the wired Internet. Unsurprisingly, DARPA chose the same distributed architecture for the wireless Internet that has proven so successful in the wired Internet. Although experts and policymakers have yet to agree on all the aspects of 4G wireless, two characteristics have emerged as possible components: end-to-end Internet protocol (IP) and peer-to-peer networking. An all-IP wireless network, which is simply a network that handles the exchange of information between users exactly as the wired Internet does, makes sense because consumers will want to use the same applications they are already using in wired networks. So wireless users will have an experience that is identical to wired users. A peer-to-peer network, where every device is both a transceiver (both a *trans*mitter and a re*ceiver*) and a router/repeater for other devices in the network, eliminates the spoke-and-hub weakness of cellular architectures, because the elimination of a single node does not disable the network, just as it does not disable the wired Internet. So wireless networks will have the same reliability as wired networks. The final definition of 4G will have to include something as simple as this: if a consumer can do it at home or in the office while wired to the Internet, that consumer must be able to do it wirelessly in a fully mobile environment. Your office will equal your home office, which will equal your mobile office. You will connect and it won't matter where you are—all the applications you normally use will be available for your use.

Figure 2.1 Example of a Peer-to-Peer Network[9]

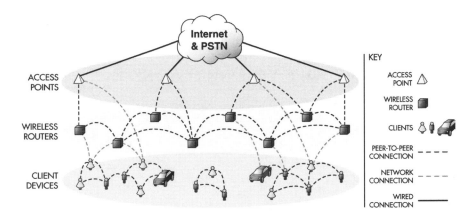

Figure 2.1 shows how to extend the public switched telephone network (PSTN, i.e., traditional phone lines managed by switches) and the wired Internet (i.e., digital traffic that moves in packets through routers) using wireless technology. In simple terms, peer-to-peer wireless networks can be created ad hoc and spontaneously—using only a cluster of users—in addition to the more traditional, planned cellular networks. The range between users would depend on many factors, mostly if the devices had line-of-sight to each other. They could all talk to each other, but could connect to the World Wide Web only if one of the users was connected to Internet, either wired or wirelessly.

If 4G—or 5G, depending on the time frame and the marketing spin that MNOs use—ends up being a wireless ad hoc peer-to-peer network, the consumer will see tremendous advantages over the current 802.11 Wi-Fi

---

9   Graphic courtesy of Eric Love. Reprinted with permission.

wireless hot spots and cellular networks. Peer-to-peer networking is significant because users joining the network add mobile routers to the network infrastructure. Because users carry much of the network with them, network capacity and coverage is dynamically shifted to accommodate changing user patterns. Simply put, cellular towers are built where MNOs think people will be. Peer-to-peer networks exist where people actually are. As people congregate and create pockets of high demand, they also create additional routes for each other, thus enabling additional access to network capacity. Certainly peer-to-peer networks will not replace traditional cellular networks in rural or other sparsely populated areas.

This technology has not been built into mobile phones yet, but is available to U.S. military units and police and fire departments. For the military, peer-to-peer networks mean no towers to try to erect in the middle of a battlefield. For police and fire fighters, peer-to-peer networks mean a way to communicate after a natural disaster or act of terrorism destroys the existing infrastructure. For users, peer-to-peer networks, when available, will mean we will automatically hop away from congested routes to less congested routes (load balancing). This will permit the network to dynamically and automatically self-balance capacity and increase network utilization. What may not be obvious is that when user devices act as routers, these devices are actually part of the network infrastructure. So instead of MNOs subsidizing the cost of user devices, consumers actually subsidize and help deploy the network for the MNO. The bottom line is that peer-to-peer mobile networks offer the promise of lower cost, higher bandwidth, and the kind of reliability we get with wired networks.

MOVING BEYOND CELLULAR NETWORKS Cellular networks have served us well. But as the number of subscribers increases—coupled with the increased demand for mobile broadband data and not just voice service—we will eventually max out the capability of cellular networks. First responders—police, fire, EMS—are already using mesh networks so they won't have to compete with the citizenry during emergencies. The next generation of mesh networks will allow businesses to connect directly to their customers, taking apart many of the barriers that exist today. Imagine products in a store *talking* to consumers' mobile phones as they walk by. Or our refrigerators talking to the contents therein to make sure the milk is not past its expiration date. Peer-to-peer networking makes all this possible.

Peer-to-peer wireless networks will be, at the very least, a component of future wireless networks because they offer such an elegant solution to the problems of expanding existing cellular networks. Less clear is what will become of our cellular networks. Should HSDPA networks—those 3.5 networks now in Japan and South Korea that deliver much higher bandwidth than even some wired networks in the United States—continue to do well in Japan and Korea, that technology will likely dominate the Asian region. Thus if the early adopters in Asia shun peer-to-peer networks, the costs of the infrastructure components will not reach economies of scale as quickly, raising the cost of such networks, and further slowing their deployment.

In South Korea, for example, the government has also licensed MNOs to offer mobile WiMAX (also known as 802.16e and wireless broadband [WiBro] networks). WiMAX is a tower-based system, but is not a cellular technology. Originally conceived as a way to connect homes to the Internet, in addition to DSL and cable, newer versions will also support mobility, which DSL and cable cannot. The U.S. company Sprint has done limited trials of mobile WiMAX in the United States. It is possible—maybe likely—that WiMAX will be more successful in the United States than in Asia precisely because our cellular networks are so far behind those in Asia. Americans may need WiMAX to get the kind of data rates Asians will get from HSDPA.

This alphabet soup of competing technologies is moving us toward the holy grail of wireless: broadband data rates while mobile. In figure 2.2, the upper-right-hand quadrant is the best of both worlds today: speed and mobility.

Figure 2.2 Comparison of Wireless Technologies[10]

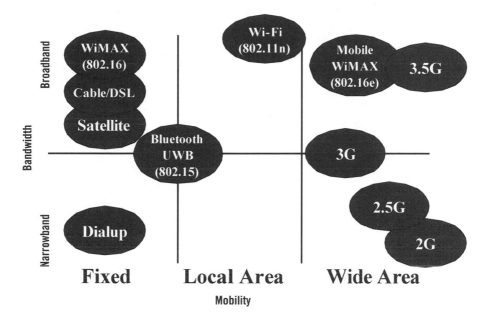

While all these technologies compete against each other for consumer dollars, they also are working together toward the migration from cellular and other "star-shaped" wireless networks (e.g., traditional cellular or 802.11/ Wi-Fi) to those that more closely resemble the wired Internet, often called peer-to-peer or mesh networks. Note that all cellular networks are star shaped with a tower in the middle, but not all star-shaped networks are cellular. For example, 802.11 Wi-Fi hot spots have a router in the middle and don't support mobility or handoffs like cellular networks do. Since the type of the wireless technology affects the speed and reliability of the wireless network, it's worth understanding the differences between these technologies.

10 Graphic created by the author, August 2007.

STAR-SHAPED VS. PEER-TO-PEER

By 2010, only 20 percent of DoCoMo's users will be human beings. The rest will be cars, bicycles, computers, ships, vending machines, and home appliances.[11] —**DR. KEIJI TACHIKAWA, CEO, NTT DOCOMO (2004)**

The cellular networks that virtually everyone in the United States is using today to connect their mobile devices to the towers in the center of the closest cell are a type of star-shaped networks. Star-shaped networks allow MNOs or 802.11 Wi-Fi hot spot operators to build out the coverage for whole cities piece by piece. Peer-to-peer wireless networks, usually called mesh networks, almost certainly have towers as part of their overall infrastructure. But peer-to-peer wireless networks don't require this central transceiver point because a user can connect to the Internet via another user who has connectivity via a tower. A peer-to-peer wireless network can do everything a star-shaped network can do, but it also provides the added benefits of supporting peer-to-peer communications. Wireless peer-to-peer networks are more resilient and flexible, and they don't necessarily experience all the problems star-shaped networks do during periods of high demand.

Furthering the economic argument away from traditional star-shaped to wireless peer-to-peer networks is the 80/20 rule. With traditional wireless networks, about 80 percent of the cost goes to site acquisition and installation, and just 20 percent is for the technology. With wireless peer-to-peer networking, however, about 80 percent of the cost is the technology and only 20 percent is the installation. Because technology costs tend to decline over time, wireless peer-to-peer networks will, over time, be much less expensive.

With a cellular infrastructure, users contribute nothing to the network. They are just consumers competing for resources. But in wireless peer-to-peer networks, users cooperate—rather than compete—for network resources. Thus, as the service gains popularity and the number of users increases, service likewise improves for all users. And *users* doesn't just mean people—it will mean cars, traffic lights, and other objects, private and public, mobile and stationary. A car in an accident can send an alert to emergency services; a drawbridge can detect ships coming in; a lost pet can phone home.

---

11   Retrieved May 12, 2008, from http://www.japaninc.com/article.php?articleID=1387.

Just a few years ago, you needed to put a card into your laptop to connect to Wi-Fi hotspots in a coffee shop or an airport. Now laptops come with the same technology built in, transparent to the user. So too with new wireless access technology, meaning that in the coming years wireless will vanish entirely from view, as wireless transceivers are embedded in everything from vending machines to your toaster oven. These wireless transceivers, and the networks that link them together, will almost certainly prove to be the most potent wireless network ever. "Gizmos and gadgets will talk to other devices and be serviced and upgraded from afar. Sensors on buildings and bridges will run them efficiently and ensure they are safe. Wireless systems on farmland will measure temperature and humidity and control irrigation systems. Tags will certify the origins and distribution of food and the authenticity of medicines. Tiny chips on or in people's bodies will send vital signs to clinics to help keep them healthy."[12]

With all this interconnectivity, however, the security and privacy concerns of peer-to-peer wireless networks will need much more attention, especially as virtually every device running on electricity gets connected to the Internet and packets are suddenly flowing through other users' devices instead of fiber optic cables buried in the ground. Security and privacy in global networks is particularly challenging, since different governments and different cultures have dramatically different norms. Compared to someone sitting in Los Angeles, the Chinese government may have a much different view of what can be uploaded to a blog about Tibet, and the government's right to demand from the application provider the name of the users uploading that comment. Even those of us in Western democracies often assume that privacy is guaranteed by some kind of implied contract between us and a company, or between us and the government. But in a world where virtually all networks interconnect and information is widely shared, that will not work. Consumers and governments will have to rethink what security and privacy mean. Do I have an expectation of privacy when my content is in the air for anyone with a free software program to capture and read? Have you seen those warnings when you connect to 802.11 Wi-Fi hot spots in coffee houses: you are connecting to an unsecured network. But many people still use these networks to log on to financial sites. Users need to understand what

---

12 *When Everything Connects.* (2007). p. 11.

kinds of networks are secure, what kinds are private, what kinds are both, and what kinds are neither.

Security must mean that we have confidence that what we send and receive over peer-to-peer wireless networks can be read by only the sender and receiver. We have a certain level of confidence that when we send grandma a birthday card via traditional snail mail, no one will read what we have written except her. We trust the post office. Peer-to-peer wireless networks cannot offer less security or privacy than traditional wireless networks if they have any hope of being universally adopted.

PRIVACY Privacy and security issues will slow the pace of implementation of peer-to-peer networks. "What if I don't want other cars to know how fast I'm driving?" "What if I don't want someone else's Internet access to be routed through my phone at the expense of my battery life?" First responders (police, fire, EMS) are already using peer-to-peer devices internally because privacy and security are less of an issue. But peer-to-peer technology for consumers—initially certainly just for those who opt in to use it—will offer new and creative ways for viral marketing or to reach a geographically targeted audience, like fans at a football game.

THE FUTURE OF DRIVING

One cutting-edge application of peer-to-peer mobile, wireless technology is in cars. Cars offer a useful and inexpensive way to populate mesh networks because they do not need additional power and tend to be where people can use them as nodes in the network. You may have seen those commercials with Batman locking his keys in the Batmobile and calling OnStar to open his door remotely. This service also gives the Dark Knight access to personal concierge services, roadside assistance, and—in the event of a bad encounter with the Joker—can even send his medical records to the Gotham Hospital, for about $30 per month.[13] The magic of OnStar works through a combination of a cellular network, the Global Positioning System (GPS), and a staff of employees who manually process many of Batman's requests. "Holy twentieth century," Robin might say.

In the event of another Batman sequel, he might want to save Gotham and the rest of the world with a car that can talk directly to other cars, get real-time streaming audio and video, obtain geo-location services with greater accuracy than GPS, and download his favorite websites. And he'll be able to do all this at 60 mph (95 km) at download speeds faster than when he is sitting back in the Batcave with a cable modem.

The emerging market concerned with sending data to and from cars is called *telematics* and comes from the combination of *telecommunications* and *informatics*. It means those products, services, or support systems that provide information to cars and other vehicles. If the definition seems broad, perhaps it is because when you add a high-bandwidth link to every car, the list of services you make possible is also quite broad. The OnStar-type basic services are easy to deliver and one doesn't need a lot of imagination to see the utility. According to a Nikkei Business study of 20,000 Japanese drivers, the applications for telematics that drivers were most interested in were largely security and safety related. About 46 percent of the respondents wanted such a service to find missing cars; 41.3 percent selected "notification to the owner's mobile device when his/her car is operated while unattended"; and 38.2 percent chose "requesting assistance in case of emergency." Further down the scale, "prediction of an optimal travel route" was ranked tenth at 22.4 percent. "Providing maps and traffic information" was ranked thirtieth at 12 percent.[14] These last two choices are probably the most used features in the United States today via the NeverLost and similar systems in rental cars. More and more of these applications are available in Japan today.

---

13   The list of services and pricing information was retrieved May 12, 2008, from http://www.onstar.com/us_english/jsp/plans/index.jsp.

14   *Drivers Expect Telematics to Promote Security.* (2002).

CAR TRAFFIC Car-to-car, car-to–car-dealer, and car-dealer–to-car communications will make driving safer. The ability of cars to share information with other cars—traffic conditions determined by average speeds, the amount of moisture on the road ahead, and traffic light malfunctions—will happen, but not in the near future. First there will be a greatly enhanced OnStar-type system where car dealers tell cars it is time for an oil change and cars are telling car dealers that a part needs to be replaced, so make sure you have it in inventory.

One issue is, then, what would adding bandwidth capacity to this network accomplish given that so much can already be offered today? How about a service that allows you to tap into the real-time streaming local Department of Transportation cameras that are already mounted throughout the city? You are stuck in traffic and want to see how many exits you need to pass before the congestion eases. Then you can decide to wait it out or use the technology at your fingertips to find an alternative route.

How about a "where is the nearest 'fill in the blank'?" application that MNOs could provide on behalf of restaurants, hotels, and gas stations? How about comparison shopping for the best price for gasoline at a given exit? These services would provide revenue for the MNOs in terms of advertising dollars from featured businesses, while at the same time providing useful, opt-in services to consumers.

How about streaming video to the flat screen in the backseat? Or streaming audio in lieu of your current radio or satellite radio systems and their recurring monthly fees? In the simplest of terms, broadband telematics offers drivers and their passengers the same Internet experiences they can get at home.

The wireless revolution affects not only the passengers inside the car, but the ability of the cars themselves to talk to other cars, your repair shop, or computers inside your garage. Lee Bruno suggests, "With all of these smarts under the hood, a car will be able to warn its driver that its radiator is going to malfunction long before it actually happens and then provide the location and directions to a nearby repair shop."[15]

Technology to accomplish all this is already available and in many cases tested and produced, with Japan's Toyota being among the world's leaders. The main reason for the slow mass implementation of telematics in the United States is almost certainly the cost to the manufacturer. Car companies are not going to put $400 worth of equipment in a car, betting you'll order a $30-a-month service. More likely, this will be something extra you buy for your car after-market, like satellite radio.

One of the most intriguing possibilities is that the peer-to-peer wireless revolution will soon include machine-to-machine communications in addition to person-to-person and person-to-machine. Imagine that you are driving, turn left, and slam on your brakes to avoid an accident. You avoid the stopped traffic ahead, but there are lines of cars behind you, turning left, that will never be able to stop in time. The result is a massive pileup. Now imagine when you slam on your brakes that your car (not you) sends a message to all the other cars (not their drivers) in the area to slow down and they do in fact slow down without driver intervention. The accident could be avoided (see figure 2.3).

Figure 2.3 Car-to-Car Communications[16]

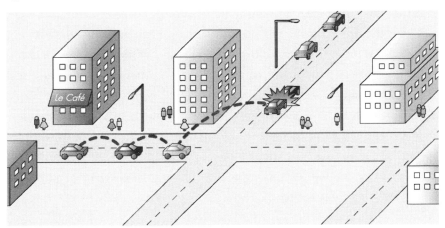

---

15   Bruno, Lee. (2002). *Building the Real Information Superhighway.* pp. 66–67.
16   Graphic courtesy of Eric Love. Reprinted with permission.

The two debates I hear the most are that this will be an expensive addition to cars and that the system would work only if every car had it installed. As to the expense, consumers would pay for it just as they now pay for anti-lock brakes, and then get a discount on their auto insurance because they are less likely to be involved in an accident. As to every car needing the system to make the de facto network effective, that is a good thing. There is a story—perhaps apocryphal—of Alexander Graham Bell asking J. P. Morgan for $100,000 to start what would become AT&T to provide telephone service. Morgan is said to have replied, "What good is a telephone unless everyone has one?" Yes, all cars will need one, thus driving down the cost of the hardware as manufacturers quickly reach economies of scale. And there will be a great after-market business to retrofit existing cars.

Peer-to-peer wireless technology is being tested on South Korea's extensive subway system after a terrible accident in 2003 in Taegu, a major city in the center of South Korea. An arsonist had set a subway train on fire in the station. A second train, not knowing what was happening in front of it, pulled into the same station and also caught on fire. More than three hundred people were killed or seriously injured—more on the second train than on the first.

In response to public demands to enhance subway safety, South Korea experimented with putting a video camera on each platform and sending real-time streaming video to subway trains approaching the station. The idea was that the driver on the train could see a problem and have time to stop before that train contributed to the problem. But traditional wireless technology made this impossible due to the speed of the train and the bandwidth required to send the video down the snaking tunnel the subway train drives through. Peer-to-peer wireless allowed for small routers to be placed along the inside of the tunnel so the video could stream from the camera through the routers to the approaching train's screen far enough from the station to give the driver time to stop the train. The next phase will be a government award to deploy this in all stations.

WILL WEB 2.0 MEAN MORE OF *LESS*?

Anything that gets invented after you're thirty is against the natural order of things and the beginning of the end of civilization as we know it until it's been around for about ten years, when it gradually turns out to be alright. —DOUGLAS ADAMS

As the wireless revolution proceeds, things are going to keep changing—fast. For some of us, that might mean playing a little catch-up. Douglas Adams is right—after thirty something changes and a lot of technology becomes harder and harder to understand. Web 2.0 is one such change that the thirty-and-over crowd were slow to comprehend, but now successful businesses are adopting—though perhaps not embracing—many Web 2.0 applications.

Web 2.0 encompasses the broad range of user-generated content (blogs, photo and video sharing), user-managed content (wikis), and social networking sites. Some service providers you may have heard of include Blogger (blogs), Flickr (photo sharing), YouTube (video sharing), Wikipedia (a free and collaborative online encyclopedia), and MySpace, Facebook, and LinkedIn (social networking sites). The key innovation of Web 2.0 has been that instead of big media conglomerates—even early dominant web players like AOL, MSN, and Yahoo—deciding what content users can see, users themselves are providing and promoting what other users see. Instead of one-to-many communication, Web 2.0 focuses on many-to-many.

The Web 2.0 revolution has had a profound impact on businesses in the United States. Customers are not just reading about products and services in the marketing collateral created by the company (one-to-many); anyone can say whatever they want about a product or service and anyone can read it (many-to-many). Web 2.0 is rapidly transitioning to a new era, one which might be called the age of personal or participatory media. Before, if I had a bad meal at your restaurant, maybe nobody knew but the waiter and anyone I happened to talk to about it before I forgot. Now, everyone can know, because I can go to any of hundreds of different sites and post my bad review. Or I can simply post a bad review on my own blog that is linked to hundreds of other sites. Businesses must now do a web search of their company's name to see what others are saying about it, and even then they can't control the messages they find.

This kind of participatory media is powerful in the wired world, but even more powerful in the wireless, mobile world. Now I can check what the world thinks about your products and services before I make a decision to buy, if I plan it out and do some research. With ubiquitous wireless networks, though, I can read those reviews in the middle of the mall or the grocery store, as I compare products. I can read a restaurant review while standing on a street corner trying to make a decision of where to eat.

Contrast this new participatory media with the early days of television in the United States, where three national broadcasters chose all the content, when you watched it, and where. If you owned a television in the mid-1950s, you almost certainly watched *I Love Lucy* every Monday night at 9:00 p.m. EST sitting in your living room.

When the time shifting that digital media allows is coupled with the place shifting that wireless networks enable, the user can choose what he wants to watch, when he wants to watch it, and where he'll be while watching it. And the choices are not limited to the 500 cable stations—anyone with a digital movie camera and an Internet connection can upload content to share with the world. Admittedly, the "cat riding a skateboard" genre of video that seems still to dominate YouTube and other video sharing sites is not quite the same as *I Love Lucy*. But now users are deciding which of the two to watch—not the networks. As Andrew Keen explains in *Against Open Culture*, "Web 2.0 start-ups are now radically undermining the closed system of traditional mainstream media."[17]

When I travel around the country speaking about *The Future of Less*, I often use the phrase *web two dot oh* rather than either of the more stylish *web two oh* or *web two point oh* because all the hype reminds me of the dot com bubble that burst in 2000 and sent the stock market tumbling. Back then we had companies mailing forty-pound (18 kg) bags of dog food to consumers (Pets.com), an alternative to cash (Flooz.com), and an online grocery (Webvan.com) going public, spending millions for Super Bowl commercials, and buying the naming rights to professional sports stadiums for up to $20 million each. "That will never happen again," we were told.

17 Keen, Andrew. (2007). *Against Open Culture*. p. 3.

Alas, Web 2.0 has seen similar silliness and unrealistic valuations. News Corp bought social networking site MySpace for $580 million in summer 2005, a brand now more closely associated with child predators than is *Dateline*'s Chris Hansen. Later that year, eBay bought VoIP service provider Skype for at least $2.6 billion, only to be forced to write off more than half of that value just two years later. And in October 2007, Microsoft invested $240 million in social networking site Facebook, valuing the company at $15 billion, more than the GDP of about 100 countries.

The bad news for the acquiring/investing firms is that any positive return on investment is seemingly years and years away. The possible good news for users of these services is that the acquiring/investing firms have deep pockets to fund the development of new applications and services. And adding mobility to social networking sites would seem to be a way for them to grow their subscriber base or at least keep more people on their site for longer periods of time. Students walking around campus cannot easily log in to Facebook, for example, but they could upload new content or reply to online visitors if a wireless interface were supported and easy to use. As wireless networks get more robust in the United States, wireless Web 2.0 will be a relatively easy application to transition to.

This marriage between more robust wireless networks and existing and new Web 2.0 applications is what we are now calling Mobile Web 2.0. It will dramatically change the World Wide Web and the mobility landscape as we know it today. The mobile web will become the dominant access method in many countries of the world, since users there will likely be using something that more resembles a mobile handset than a desktop computer. And the offspring of this marriage is evolutionary, not revolutionary. It is the natural progression of two important technologies once their paths cross.

Odds are that you have used some Web 2.0 services from your traditional desktop or laptop PC. If you have a mobile phone, it is natural that you would want to use these same applications anywhere, anytime. The under-thirty crowd that still dominates a lot of Web 2.0 content creation and consumption is the mobile generation, insistent on the immediacy that only wireless can support. For those a little older, this may seem as foreign to the cultural revolution of the 1960s and 70s that our parents endured.

Uploading a picture you just took with your camera phone to your blog, or watching a YouTube video on your mobile phone, are simple to do from a technology standpoint. But there is a lot more you can do with a mobile device. Imagine you are at a party and you see someone across the room that you might want to talk to. A few clicks on your mobile phone could reveal her age and profession, links to her latest blog posts, and a plethora of other personal information. This might sound scary (if someone were doing this to you), but most (though not all) of the content that would be accessed is what you posted about yourself. This is mobile Web 2.0. A mobile etiquette will have to arise about when it is okay to research your pre-date, just as most societies have basic rules for when a man can approach a woman at a bar and clumsily ask, "Do you come here often?'"

> **WEB 2.0** Those user-centric (participatory) Web-based applications like Facebook, YouTube, and Wikipedia may (one could argue, should) be part of your business's internal and external communications plan. But adding the mobility to these applications means new opportunities and challenges to your business. At the very least, wireless Web 2.0 (a.k.a. mobile Web 2.0) allows for place shifting—consumers can access your content where they want it, not just at home or in the office. Maybe this is happening in the middle of your grand opening, or product launch, or in the audience you are presenting to. Will this impact what kind of content you will or won't want to deliver?

The majority of today's Internet users still access it via their computers. Globally, however, accessing the Internet via a mobile device is growing much faster (as a percentage). So, as tech start-up Funambol (its name apparently from the Latin for *tightrope walker*) describes it, Mobile Web 2.0 is the outgrowth of several compelling trends:[18]

- Better, faster, cheaper wireless devices, networks, and data services and plans.

---

18   Retrieved May 12, 2008, from http://www.funambol.com/about/.

- Messaging maturation, where people are looking beyond text messaging for the next cool mobile experience.

- Emerging technology such as mobile search, presence, location-based services, and navigation.

The challenge for Mobile Web 2.0 firms, then, is that their core customers—those instant messaging, photo sharing, blogging customers—are primarily in the most developed countries where people have time to join social networks and update wikis. Are there enough millennials accessing the Web with mobile devices to make a business? And what, exactly, are they willing to pay for? Virtually all of the Web 2.0 applications they were using on the wired Internet, and will use on the wireless Internet, are free to them. They'll pay the MNO for access to those applications, and they'll pay 99 cents for a song or a ring tone, but will they pay for anything else? Consumers have been spoiled by user-generated content sites like YouTube and Flickr, the photo-sharing site purchased by Yahoo, because use of the software and access to the content is free.

Nonetheless, the optimism that still dominates Web 2.0 is equally present in Mobile Web 2.0. Content created on mobile devices will change the balance of power in the media industry. The mobile device is ideally poised to capture user-generated content "at the point of inspiration"—making it the main driver behind Web 2.0.[19]

We're back to the chicken-or-the-egg dilemma. Should MNOs build next-generation wireless networks in the hope that such networks will motivate other companies to create high-bandwidth services and applications so that the MNOs can earn a return? In other words, should MNOs push their networks on consumers? Or should Mobile Web 2.0 start-ups design new products and services that will require more bandwidth before it is available, and let consumers demand better mobile connectivity? The MNOs seem to have decided that the *if you build it, they will come* model exposes them to too much risk, so at least in the United States, consumers are seemingly going to have to lead us into the Mobile Web 2.0 revolution.

To try to understand whether the chicken or the egg will come first, it's enlightening to look at the model of the development of the personal

---

19 Jaokar, Amit, and Fish, Tony. (2006). *Mobile Web 2.0.*

computer. Microsoft developed the Windows operating system to run on a computer powered by a microprocessor made mostly by Intel. (Microsoft, like Mobile Web 2.0 firms, represents the software side of the equation. Intel equals hardware, just like the MNOs.) Microsoft and Intel became so codependent that the term *Wintel*, Windows plus Intel, entered the lexicon. Microsoft's operating systems (e.g., 3.0, 95, 98, XP, and Vista) and its Office suite of applications grew larger and larger, requiring more and more processing power. So people had to buy new computers with faster and faster Intel chips (e.g., 286, 386, 486, Pentium, and Centrino). Microsoft's new products would not run on old Intel hardware. And no one would have bought new Intel hardware if the new machine couldn't do anything more than the old machine.

Thus it is likely that wireless network capacity will grow slightly behind the need for it to grow, ensuring demand-side pressure that will keep prices from falling too fast. And wireless services and applications, including Mobile Web 2.0, will be launched as soon as there is a network on which to launch them. Either way, the result seems clear: "Every social network is going to have a mobile component," predicted Jill Aldort, an analyst at Yankee Group.[20]

CONCLUSION

It is the function of creative man to perceive and to connect
the seemingly unconnected. —**WILLIAM PLOMER**

For users, the shift from centralized (cellular/star-shaped) wireless networks toward the mesh-shaped (peer-to-peer) approach of the wired Internet is all good news. Most important, it will be transparent—you won't have to do anything differently to connect. And your monthly bill should decrease over time, since these peer-to-peer networks should reduce recurring monthly costs for the MNOs. Since users double as nodes for other users, less infrastructure may be required in some cases, lowering the capital costs of the MNOs. And

---

20   Taylor, Chris. (2006). *The Future Is in South Korea.*

decentralized networks should greatly increase reliability, since no single bottleneck or outage will keep us from accessing the network.

As the parent of two future drivers, the future of driving is really important to me and should be to you too (even if you don't have kids) since you'll be driving on the same roads as my kids. Peer-to-peer wireless networks in the form of car-to-car communications will be focused on safety, namely accident avoidance. It is ironic that what we often call the killer app is, with the future of driving, a way to have fewer people killed in car accidents.

Security and privacy—two recurring themes throughout this book— are critical to any technological revolution. Security is something that consumers demand from commercial service providers. My information must be secure, or at least I must have the feeling that it is secure, or I won't do business with you. Peer-to-peer wireless networks may have the perception of being less secure and companies offering this type of service will have to demonstrate that these networks are just as secure as the alternatives, which is a fairly low standard to beat. Privacy is something that consumers demand from both commercial service providers and the government. With good security, privacy is not much at risk from the technology the network is using. Privacy is more a matter of the policies and processes the government and companies employ. That is good news—policies and processes should be more receptive to consumer demand than what it would take to develop new products.

Mobile Web 2.0 is possible now with traditional wireless networks, but peer-to-peer networking will greatly increase its functionality, and diminish the cost to deploy and use it. Robert Metcalfe, who co-invented Ethernet, postulated that the value of a telecommunication network grows exponentially with the growth of the number of users. Simply put, one guy with a phone has no one to talk to. More phones being used adds value to every user because there are more people to call. Metcalfe's Law illustrates the value that wireless will bring to Web 2.0 because it allows even mobile users to connect to the network.

CHAPTER 3
POWER TO THE PEOPLE

It's time for the future to be on your phone.

It's time to e-mail and walk at the same time.

It's time to be online offsite.

It's time to make a call about an e-mail about an attachment.

It's time for Treo.

E-mail. Phone. Web. Organizer.

—TREO 700W SMARTPHONE MAGAZINE AD

Size matters. Since the early years of wireless, even before the cellular phone, there has been tremendous consumer pressure for smaller and smaller devices, and fewer of them. It all started in the mid-1970s with consumers being able to carry a single device, a pager. Then they got an analog cellular phone and carried that too. They had to carry both since the analog phone couldn't receive any data, like text messages. Beginning around 1990, digital cell phones allowed people to replace those two devices with one. Then in 1996 they got personal digital assistants (PDAs), like the Palm Pilot, and were back to having to carry two devices.

With consumer pressure to reduce the number of devices we need to carry from two back to one, the choice was either to add data to the phone or add voice to the PDA. Data on a phone was a problem—the screen was too small and the keyboard interface was not well suited for text (like having to press the 2 key three times to get a *c*). Adding voice to a PDA was much easier, but the bulkier PDA wasn't as cool as the previously ever-shrinking phones. Functionality won out over coolness and PDA phones proliferated. But the growing popularity of the iPod resulted in many people carrying two devices again. Now the iPhone and the broader movement of many manufacturers to integrate MP3 players into mobile phones allow people to carry just one handheld device again. But we are still carrying our car keys and wallet too—for now, at least.

HARDWARE CONVERGENCE

One ring to rule them all, one ring to find them;
One ring to bring them all, and in the darkness bind them.

—J.R.R. TOLKIEN, *THE FELLOWSHIP OF THE RING*

The term *mobile phone* no longer seems fully descriptive—*mobile device* is probably more accurate, if less specific. Over time we'll settle on a better term that encompasses a product that's now tasked with being our calendar, camera, contact list, web and e-mail access device, game box, photo album, and music and video library. Those who have a Blackberry almost certainly spend more time sending, receiving, and reading e-mail on it than they do talking. Most iPhone users almost certainly listen to music and watch video on it as much as or more than they talk. The phone function—that is, the ability to talk to other people—seems like a steering wheel on a car—it has to be there, but it isn't the reason you chose one model over another.

Just as the pager, PDA, and MP3 player morphed into a single device that can also make phone calls, the modern mobile phone will continue to add functionality until it becomes the Swiss Army knife of communications devices (see figure 3.1). The overwhelming majority of digital cameras sold every year are in mobile phones. We can watch television on our mobile phones. They eliminate the need for a wristwatch. And soon, they will become our e-wallet for transactions that formerly required cash, credit, or a subway token.

Figure 3.1 The Mobile Phone as the Swiss Army Knife for a Digital Life[21]

MULTIPURPOSE DEVICE A Swiss Army knifelike device is about much more than just convenience. Yes, it will be better to carry one tool that can perform many tasks, and yes, you'll know you'll always have a camera with you, always have a means to connect to the Internet, and can forego a wristwatch if you so desire—the clock on your mobile phone even updates automatically for daylight savings time. But it will be much more than just convenient. Hardware convergence means increased mobility—one handheld device to carry instead of a purse or backpack full of devices. And that means it is more likely to be in the palm of your hand, more likely to be used, and more likely to be available as a node for a peer-to-peer network. Wristwatches seem unlikely to disappear from the marketplace for fashion—not technological—reasons, but single-function digital cameras might. Is your company making horse shoes instead of cars, slide rules instead of calculators?

---

21 Graphic owned by the author. *Swiss Army* is a registered trademark of Victorinox. Other products pictured are for illustrative purposes only.

These all-in-one devices are critical components in the broader ubiquitous network family of products because their existence will drive demand for ubiquitous networks. After all, what is the point of being able to connect to any site from anywhere if I don't have a device that enables a mobile Internet experience that is almost identical to my non-mobile Internet activities? But mobile phones are still not as easy to use as personal computers. Nor are they as easy to develop applications for—unlike personal computers, where one operating system dominates, there are at least six major mobile phone operating systems.

Applications, usually in the form of software, drive the sales of hardware, not vice versa. No one buys a computer and then says, "I think I want to start sending e-mails or surfing the Web." They first say, "I think I want to start sending e-mails and surfing the Web," then buy the hardware that makes that possible. Once they own that hardware, over time they start to find and use all kinds of other applications—PC games, instant messaging, tax preparation software, and document editing software, to name a few.

South Korea has long been a leader in multifunction devices, and Samsung, LG, and Pantech are constantly rolling out new models. Years ago, Koreans were taking their PDA-shaped smart phones and dropping them into a cradle on their car dashboard where GPS maps and traffic information were automatically displayed. The combination of experienced, sophisticated wireless consumers and great wireless networks makes South Korea a useful test bed for non-Korean companies too. Motorola has often sent prototypes to Korea with many features to see what customers over there would want and use. Then it could remove the unpopular ones and go to a broader market with a core set of features. After launching the Razr2 in Korea, Motorola Korea announced, "We're releasing our new phone in the Korean market first in recognition of tech-savvy and fashion-aware Korean consumers."[22]

This will be true with the next generation of multifunction mobile devices too. You bought your first mobile phone to make phone calls. You bought your first mobile device because you wanted to send and receive e-mails or surf the Web for the latest stock quotes or sports scores. But once you own a device, you start to find additional things to do with it. So even if you didn't buy it for these reasons, your mobile phone will become, over time,

---

22 Gardner, W. David. (2007, July 3). South Korea First to Dial Motorola's iPhone Challenger.

your wallet and ATM, your house and car keys, and your restaurant critic and concierge. According to IBM, "New technology will allow you to snap a picture of someone wearing an outfit you want and will automatically search the Web to find the designer and the nearest shops that carry that outfit. You can then see what that outfit would look like on your personal avatar—a 3-D representation of you—right on your phone, and ask your friends, in different locations, to check it out online and give their opinion."[23] Mobile technology weakens the divide between the vast reserves of collective knowledge stored on the Internet and our own lives, and it does it in real time, wherever we want or need it.

In 2008, South Korean handset manufacturer Pantech held a competition among university students to create concept phones (those not necessarily destined to become full production models) for release in 2010. While many of these phones focus on form rather than functionality, it is noteworthy to see how far ahead Korean wireless handset manufacturing firms are looking.

What will the next revolutionary function be for your Swiss Army knife-like mobile device? It is something you'll never see: a cognitive radio, which will allow our devices to make decisions about how to communicate with other devices and the network. Not moving very fast and in need of a high data rate? The device will connect to an 802.11 network, those Wi-Fi hot spots found in Starbucks and airport waiting areas. In the subway and trying to check the sports scores? The device will connect to a traditional cellular network. Which cellular network? The device will decide based on an evaluation of the signal strength of all available networks.

Of course, these intelligent, empowering devices can also expose their users to the possibility of being tracked by their service providers, the government, or others with the requisite technology. The positive spin is that MNOs could offer opt-in geo-specific and time-specific marketing text messages. Round a corner at noon and your phone could let you know that a new restaurant has opened nearby. Pass that same corner three hours later and be advised that there is a sale down the block. On the negative side, warns John Gage, chief researcher and vice president of the science office for Sun Microsystems, "New cell technology automatically reports

---

23 Vail, Michael. (2007). *IBM Reveals Five Innovations That Will Change Our Lives over the Next Five Years.*

sub-meter location accuracy that could create a logistics revolution or lay the foundation for a police state."[24] Do we want technology that allows the government to know exactly where we are at all times? What will the trade-off be between functionality and privacy? And who will decide the appropriate balance?

Law and order advocates will probably argue that this is no problem. Whether or not you subscribe to the philosophy that "if I'm not doing anything wrong, I have nothing to hide," these advances raise security and privacy issues that will have to be addressed. If it is possible to know where you are, then it is possible to know exactly where you are not—when you are not at home, or not at work, or not in the country. This could be very useful information for burglars or just annoying bosses. From a privacy perspective, why would I want anyone—maybe even including my family—to know where I am all the time? The quick fix is to turn off your mobile phone. But why should you have to stop using a service that you are paying for just to prevent others from using it against you?

There are societal and legal issues too. Should husbands and wives be allowed to track the location of their spouses? Should parents be allowed to track their teenage children? Or their aging parents? And who gets to decide this? The government? The trackers? The tracked?

It is almost a certainty that you are doing things on your computer today—e-commerce, e-banking, voice and video conferencing—that you never thought of, or perhaps weren't even possible, when you bought your first computer. It will be the same with mobile devices—your mobile phone will impact many aspects of your life in ways you never imagined.

What is next for hardware convergence? The past gives us a hint about the future. There was a time in the United States, not so long ago, when there was only one phone company and it owned all the phones, even the ones in your home. You rented them every month and that charge was added to your bill. The notion was that the phone network was so important that you couldn't trust consumers to plug just any device into it. What was really the case was a monopoly exercising monopoly control.

When finally given the option of buying their own phones instead of renting them, customers overwhelmingly chose to do so. Many companies

---

24 Ibid.

started to manufacture phones and three consequences of competition quickly kicked in: lower prices, better product quality, and many more consumer choices. Mobile phones, while not rented, will follow a similar path. As the vast collection of previously stand-alone gadgets morph to create one powerful all-in-one mobile device, it will mean your new mobile phone may cost more than you are used to paying, but probably less than the cost of buying all the individual gadgets individually. An iPhone is less expensive than the combined cost of a phone, video player, and iPod. Multifunction mobile phones, therefore, may not need to be subsidized by your MNO (which MNOs now do almost universally, in exchange for a two-year agreement) because they are already cheaper for consumers than buying all the individual components like a separate camera, GPS device, and wristwatch. Even if a wristwatch today may convey a certain level of status, or serve as a fashion accessory, or just show off your conspicuous consumption, your new state-of-the-art mobile phone will similarly accomplish many of these cultural notions too. And of course, no one is stopping you from still wearing a watch.

CAPITALIZING ON HARDWARE CONVERGENCE We can see the success today of hardware convergence. And we can look toward Asia to understand where our own trend lines will take us. We are on the path laid out for us by the wireless revolutions that are underway. The challenge for business is to capitalize on those taking the path. Not many gold miners made it rich in 1849 in California. The money was made by those offering goods and services to the miners.

I've given up asking, where is my flying car? It won't happen in my lifetime and probably not in my children's lifetimes either. But where is my Swiss Army knife-like, multifunctional, all-in-one, peer-to-peer mobile phone? It will be parked at my home long before a flying car is. Why isn't such a device already here?

First, we need global standards so that devices can talk to each other. Imagine if some electrical devices plugged into only some outlets in your house and other devices required other outlets. Standardization—in everything from the size and shape of CDs and DVDs to the dimensions of phone and Internet jacks—lowers costs over time since users need only one device to interface with standards-based products. But the lack of standards has plagued the wireless industry from the beginning, and it is consumers who have suffered the consequences. Your Verizon or Sprint mobile phone using the proprietary Code Division Multiple Access (CDMA) wireless transmission technology will work only in a handful of countries. Your AT&T mobile phone using the proprietary Global System for Mobile communications (GSM) wireless transmission technology—the most prevalent technology in the world—won't work in Japan or South Korea. In fact, the overwhelming majority of mobile phones in use in the world today won't work in Japan and South Korea, despite their advanced wireless networks, because of those countries' choice of nonstandard wireless technology. Wideband-CDMA (W-CDMA), a 3G wireless transmission technology standard, offered promise of being the world standard, but now we also have some small deployments using the CDMA2000 standard. Imagine if your computer could send e-mail only to a handful of other users, instead of anyone with an Internet connection. That is what the mobile phone industry is similar to today. These nonstandard standards raise costs and delay the rollout of new networks and applications.

We'll need standards so that devices can connect to networks and talk to each other peer-to-peer anywhere in the world. Not standards imposed on us by governments or product vendors, but standards based on customer use and preferences. As a start, anyone's mobile device should be able to work in every capital city anywhere in the world. There are only a couple of protocols and a handful of frequencies. It simply is not that hard. But it is not yet clear who can take a leadership position to drive a single global standard for mobile networks, just as we have for e-mail.

THE FUTURE OF MUSIC AND VIDEO

Wireless will become the most formidable music platform
on the planet.[25] —**EDGAR BRONFMAN**

You pull up to the light, and the car next to you is playing a song you wish
you could hear from the beginning. So you pull out your mobile device, type
in the title or the artist's name, and a few seconds later, you are listening to
that song. The song was delivered to you from one of a handful of free online
radio stations or music stores. Just as CDs replaced cassette tapes and the iPod
and other digital music players are replacing CDs, music on demand will
replace everything that came before it.

> In 2006 EMI, the world's fourth-biggest recorded-music company,
> invited some teenagers into its headquarters in London to talk to its
> top managers about their listening habits. At the end of the session
> the EMI bosses thanked them for their comments and told them to
> help themselves to a big pile of CDs sitting on a table. But none of
> the teens took any of the CDs, even though they were free. "That was
> the moment we realized the game was completely up," says a person
> who was there.[26]

EMI and other music companies lost the game because of the popularity of the Moving Pictures Experts Group Audio Layer 3 file, more conveniently called MP3. It allows for digital audio files to be compressed,
while still maintaining close to their original sound quality. It's a technology
that has led to the creation of new companies, new devices, and enough
litigation to keep a whole generation of lawyers employed. The popularity of the MP3 format exploded in the 1990s primarily because it allowed
bandwidth-limited users (most people were still using dial-up to access the
Internet back then) to share music.

MP3s are portable because the compression reduces storage needs. But
a digital audio/video device still has a finite amount of memory (in the MP3

---

25 Ahonen, Tomi, and O'Reilly, Jim. (2007). *Digital Korea*. p. 160.
26 *From Major to Minor*. (2008).

format, one megabyte [MB] is roughly equivalent to one minute of music). So there is only so much music you can download to your PC and then upload into your audio/video player to take with you. Advances in small hard drives and flash memory greatly increased the volume of music the average person could carry with him. And the introduction of Apple's iTunes store, which sells audio files in the proprietary MP4 format, meant that many more songs were available to download to one's computer and then upload to one's iPod or other digital music player. While downloading direct to the iPod is possible, most people do not. Better wireless interfaces are possible, and Microsoft's second-generation Zune digital audio/video player may make music- and video-on-demand much more a reality because of this. But all popular devices still rely on hard drives or flash memory, which constitutes a considerable amount of the expense of the device. And it is not necessary. The physical limitation and cost of conventional handheld digital audio/video players can be overcome by disengaging from the download/upload process. Streaming digital music or video from network-based storage or one's home computer can supplement or even replace the local storage of current-generation digital audio/video devices (see figure 3.2). By freeing users from the limited capacity of their local storage, streaming makes available an infinite amount of music and video.

THE MUSIC INDUSTRY The iPod, the iPhone, and the iTunes music store were all important in the way we buy and listen to music. They revolutionized how the music industry worked and the iPhone helped push us down the path toward hardware convergence. Since Apple didn't invent the .mp3 music format or the .mp3 music player, there are some important lessons to learn about how and why some products are so successful and others are not. Want to read a lot more about that? Check out Pip Coburn's 2006 book, *The Change Function: Why Some Technologies Take Off and Others Crash and Burn.*

Figure 3.2 Example of a Peer-to-Peer Digital Audio Network[27]

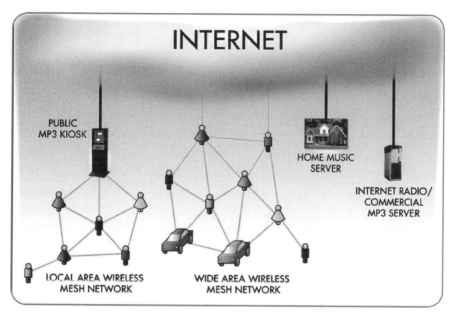

This network architecture is based on three complementary networks working together—the wired Internet, traditional wireless networks, and peer-to-peer wireless networks. The wired Internet will support the ability to store digital audio/video files on a home server, which is basically just a computer with a big hard drive that can be accessed remotely via the wired Internet or a wireless network that connects to the wired Internet. The price of home servers has been falling quickly. You can now buy a 1-terabyte (thousand-gigabyte) home server for about $700; that is twenty-five times the storage capacity of a good iPod, but only three times the price.[28] This home server can then stream the music stored on it to your mobile phone (or cars, laptops, or other mobile devices) via a traditional wireless network.

The third network that could be used in the future of music and video would be the peer-to-peer wireless networks described previously. With peer-to-peer networks, you could share the music on your device directly

---

27 Graphic courtesy of Eric Love. Reprinted with permission.
28 Wildstrom, Stephen. (2007). *Microsoft's Nifty Digital Shoebox.*

with other users nearby, subject to the constraints of the digital rights management (DRM) scheme being used. With iPod competitor Zune 2.0 from Microsoft, for example, you can send a song to a friend wirelessly and he can listen to it three times before it disappears. Because of this technology, most digital audio/video players in the future will be wirelessly enabled. Of course, stand-alone digital audio players will almost certainly be replaced by multifunction mobile devices that do many more things, such as the iPhone and other smart phones.

The recording industry was very slow to embrace online digital music sales, which resulted in many consumers turning to illegal music downloads. The recording industry responded with lawsuits and achieved some limited success in shutting down early music file sharing sites such as Napster, Kazaa, and Limewire.

However, the audio and video recording industries seem to be starting to understand that they have much to gain from this peer-to-peer wireless model if they begin to embrace a direct business-to-consumer distribution model. For example, why should EMI license its music to Apple's iTunes store instead of just selling their music inventory directly to the end user? Record companies could sell subscriptions, entire albums, or individual songs. This would substantially reduce music and video companies' production, distribution, and inventory costs, and thereby considerably improve margins and profitability, or allow these companies to lower the price per song. The tremendous success of iTunes shows that people will pay for digital music if they can pick and choose the content.

New distribution channels are also possible with peer-to-peer wireless networks. For example, peer-to-peer technology would allow for a revolutionary consumer-to-consumer distribution channel. Imagine if you could download a song or movie and let me hear it or watch it on your mobile device. If I also wanted to have it on my device, you could send me that file directly, peer-to-peer, from your mobile device to mine. I couldn't listen or watch it until I paid for it from the same vendor that you bought it from. And that vendor would know I got the file from you so you would get credit for that sale to me, perhaps in the form of a discount on a future purchase. So current customers would become a particularly good distribution channel, given that music and movie sales are often driven by recommendations from friends. These same customers will essentially become the equivalent of marketing staff working on commission for the music and video companies. And music and video content owners will have yet another way to sell their products to consumers to offset the continued decline of CD and DVD sales.

CONCLUSION

If you don't change the path you are on, you are likely to end up where you are heading. —**ZEN PROVERB**

Three components of the wireless revolution—the migration from wired to wireless, the shift from star-shaped wireless networks toward something more resembling the peer-to-peer approach of the wired Internet, and the transition from powerful networks to more powerful devices—all exist today in one form or another. The first, wired to wireless, is now so much a part of our culture that there is no turning back. Don't invest in the stock of pay-phone companies.

The second component—the deployment of peer-to-peer networks—is likely to occur, but less certain. Some experts believe that such networks will be required to achieve the kind of mobile broadband data rates that future applications will require. But with 3.5 HSDPA and the possible success of WiMAX, what we think we know about the limitations of star-shaped networks may change.

The third component of the wireless revolution—powerful devices created by hardware convergence—is all but inevitable and already rapidly shaping our culture too. What hardware convergence means is that devices that do not promote convergence—single-function MP3 players, dedicated e-book readers, and handheld GPS devices come to mind—will not be successful long-term. As a business strategy, it would be better if those firms offering such products instead licensed their intellectual property to manufacturers of multifunction consumer electronic platforms. MP3 players have already largely made a place for themselves inside mobile phones. A high-resolution screen, like that of Amazon's Kindle e-book reader, might make mobile phones easier to read outdoors. And the accuracy and user-friendly interface of some GPS devices would certainly add value to mobile phones.

Also less certain is exactly what we will be using our wireless devices to do. *BusinessWeek* tech guru Stephen Wildstrom argues,

> It is hard for anyone to picture how wireless will be used (in the future), just as it was with electric motors and microprocessors. Wireless technology will become a part of objects in the next 50 years (just) as electric motors appeared in everything from eggbeaters to elevators in the first half of the 20th century, and computers colonized all kinds of machinery from cars to coffee machines in the second half. Occasionally, the results will be frightening; more often, they will be amazingly useful.[29]

That is the beauty to me of the wireless revolutions—they will enable "amazingly useful" applications that we cannot even envision yet.

---

29 Ibid.

SECTION 2

THE FUTURE
OF PAPERLESS

CHAPTER 4
PAPERLESS DOESN'T MEAN NO PAPER

Nature provides a free lunch, but only if we control our appetites.

—WILLIAM RUCKELSHAUS

Of the three revolutions I describe in this book, the paperless one is both the easiest and hardest to envision. Wireless means the mobile phone we carry everywhere. Cashless means the credit and debit cards in our wallet. What paperless means to me is that all written content should be available in a digital—thus paperless—format. For those who want to hold what they are reading, they can print it out for themselves. But the default will be paperless since it is easier to go from bits to atoms than it is to go from atoms to bits.

We're already sending billions more paperless e-mails than we are letters. But the wireless revolution was inspired by our need for mobility and the convenience that results, so we gladly sign two-year service agreements and pay a lot of money for our wireless service. Cashless exploded onto the scene because it came with credit. We were suddenly able to spend as much as we needed because we didn't have to trade pieces of paper for it. In contrast, going paperless seems like something we have to do, not something we want to do.

The path toward a paperless society may require a Moses-like forty-year trek through the desert. The generation that brought us computers and the volume of paper they produce might have to be replaced by a generation that better understands how computers and the Internet can actually reduce paper consumption. The *Zits* newspaper cartoon features Jeremy Duncan, a fifteen-year-old living with technologically challenged parents. One of my favorites has Jeremy leaning over his father's shoulder while the latter is reading the daily newspaper. After reading for a frame or two, Jeremy walks away with a confused look on his face, saying, "Why are you reading that? All of that stuff happened *yesterday*." For everything that paperless is and isn't, digital content is and always will be a faster way to disseminate ideas than paper. I used to read *Zits* in my local paper; now it arrives seven days a week in my inbox for free.

Most of you are probably reading this on paper in a traditional book, with a cover, page numbers, and no hard drive or flash memory. You like books. So do I. I take a book to the beach, not a laptop or an e-reader. I still get a daily newspaper and occasionally write an article for it. I subscribe to about six paper magazines. As I tell my professional colleagues, paperless doesn't mean no paper.

The heavily hyped paperless office never materialized either. Many of my colleagues (and probably yours too) still print office memos and put them in my mailbox.

If we can agree on what paperless is not, what is it? To me, going paperless is a conscious personal or business management decision to reduce paper usage and enjoy all the benefits that going paperless brings. It means simply that technology is giving us options to reduce or eliminate much of the paper we use today so as to enjoy the increased portability of ideas that are digitized, reduce the costs of transmitting information, and promote a greener way of doing business.

THE INCREASED PORTABILITY OF IDEAS

Paper is no longer a big part of my day. I get 90 percent of my news online, and when I go to a meeting and to jot things down, I bring a Tablet PC. It has a note-taking piece of software called OneNote, so all my notes are in digital form.[30] —**BILL GATES**

My fifteen-year-old son has terrible handwriting, something he likely inherited from his father. I spent many hours of my own childhood—against my will—trying unsuccessfully to improve my handwriting. To his face, I encourage my son to take his time and to try to print more clearly so as to please his teachers. Privately, I lament that so much time is being spent on a skill that will end up in the dust bin of history next to the slide rule.

This is not to suggest that people will stop writing, but rather that they will be able to stop writing in longhand if they so choose. The advent of the calculator didn't mean that slide rules were outlawed, but simply that there was a choice in the marketplace, and, over time, the newer product won greater acceptance because it served needs that the older technology did not.

So too with the migration from paper to digital content. People will still have printers on their desks to change their bits into atoms, bring hard-copy notes to meetings, and later file them away in manila folders. But if they so choose, they will be able to enjoy the cost savings, storage ease, distribution options, and eco-friendly aspects of going paperless.

There are dozens of websites in this book that, if you were reading it in a digital format, could be links that would take you somewhere else with the click of a mouse. Did you read a paragraph that you liked and wanted to share with a colleague? Cut and paste is so much easier than scanning the page into a .pdf or retyping the text. Was this book expensive? Might it have been less without the paper, the printing company, the warehouse, the bookstore, and the shipping or sales tax? Could the author have sold it to you for less than you paid, while still earning more for himself? That's paperless.

---

30   Gates, Bill. (2006). *How I Work.*

SHARING DATA Let's say there is a graph in this book, or a couple of paragraphs that you'd like to share with a colleague. What are you going to do? Squish this down as flat as you can on a photocopier or a scanner and get a copy or an image that can be shared? Hunt through your bookshelves the next time you host a party to show off the figure you like? Wouldn't it be easier if you could get all content in digital format and then convert it to atoms only when you need or want to? Paperless means making content easier to share.

Paperless doesn't mean the end to books or magazines or newspapers or children's coloring books or family photo albums. Paperless doesn't mean the end to business cards, wall calendars, diplomas, maps, or manila folders. It doesn't mean my children and grandchildren won't have to learn and practice their handwriting.

What has changed is that content that was until recently only in paper form—newspapers, magazines, books, photos, etc.—will also be digitized for easy and timely sharing. Before, when I came across an interesting article in a magazine, I used to cut it out and use snail mail to send it to my father. I still read most magazines in their paper format, but now I go online and send the digital version of the article to my father, who gets it while it is more relevant. I also saved a stamp, an envelope, and a trip to the post office. I didn't save a tree, but I probably did less harm than I would have the old way. And if everyone did it this way, then certainly we would save some trees. "Stack your pennies high enough and you get dollars," my grandmother taught me growing up. There is a cumulative value in going paperless. I bought my father his first computer about ten years ago. Previously he would type a letter on his manual typewriter, photocopy it twice, and mail it to me and my two sisters. Now he opens Outlook, selects the "kids" group I created for him with our e-mail addresses, writes a letter, and hits send. I get about four paperless letters a week now instead of one, and the news is more current.

To me, paperless means digital content that is easy to share. And it's free, or at least the incremental cost of sharing it is zero. Studies show that people read around 10 MB worth of material a day, hear 400 MB a day, and see 1 MB of information every second.[31] Can you imagine printing out 1 MB of information every second? Paper is not an efficient medium to exchange information. We don't cut letters in big stones or paint pictures on cave walls anymore either.

I think many people cannot even envision a paperless world. When I asked, "When you left the house this morning, what did you carry?" probably you gave answers that relate to wireless (your mobile phone) and cashless (your purse or wallet with a credit and debit card or two). You didn't say anything paperless.

You are, however, far more paperless than you were a few years ago. You're sending more e-mails and fewer letters. You've probably sent a few e-cards to friends and therefore made fewer trips to the Hallmark store. You read an article online rather than in a magazine. The wireless and cashless revolutions are obvious and are now intricate and critical parts of our lives. Paperless is more subtle. Paperless is not what we carried this morning. But we are carrying less every day whether we know it or not.

THINKING BEYOND PAPER So you're reading my ideas about the paperless revolution in a book. If you like them, please don't feel like a hypocrite: paperless doesn't mean no paper. Books, newspapers, and magazines are not going away anytime soon. When you design new marketing campaigns, however, or new internal communications programs, you must think beyond paper. How closely do you peruse the direct mailings that fill your traditional mailbox? Most likely, you throw them away unopened. But just as likely, you delete the electronic version of direct mail in your e-mail inbox too. Simply going paperless is not enough to reach your customers. It must be part of a larger marketing strategy.

---

31 *The Phone of the Future.* (2006). p. 18.

The notion that the trend toward paperless matters less than wireless may change. Just as many are now addicted to their cell phones and cannot imagine life without them, so too may we get hooked on paperless perks—reading our local paper online when we are out of town; books on demand; remote, online access to what were formerly paper files in a cabinet in our office that otherwise would have required a trip in a car to view. The Internet and wireless revolution give us more reasons to start on the path toward paperless.

THE COST SAVINGS

Even a sheet of paper has two sides. —**JAPANESE PROVERB**

Technology is giving us options to reduce or eliminate much of the paper we use today so as to enjoy the cost savings of being paperless. "The most visible impact of a move to a paperless office is the reduction in the cost of printing, mailing, shipping and storing paper," notes Joseph Anthony, author of the very useful *Six Tips for a "Paperless" Office*. "People who have been making photocopies, sending paper faxes, putting documents into legal size folders—or saving mounds of mail and catalogues that they just can't part with—are going to have to change their perceptions."[32] If you need further motivation, a survey by furniture producer Steelcase found that 30 percent of white-collar workers still have private offices, but the size has shrunk from an average of 16 by 20 feet a few years ago to 8 by 10 feet today.[33] Getting rid of all those filing cabinets might make eating lunch at your desk a little less stressful.

Take a look around your own company's office supply room. What do you see that could be eliminated if your firm were to adopt a path toward paperless? In other words, what would you no longer have to buy, or could you at least buy less of, to enjoy the savings from that policy?

---

32   Anthony, Joseph. (n.d.). *Six Tips for a "Paperless" Office.*
33   Woyke, Elizabeth. (2007). *Wanted: A Clutter Cutter.* p. 12.

Copy Machines	Legal Pads	Staplers
Copy Paper	Paper Clips	Staples
Envelopes	Pencils	Stationery
Erasers	Pens	Sticky Notes
File Folders	Printer Cartridges	Tape
Filing Cabinets	Printer Paper	Tape Dispensers
Highlighters	Printers	Three-Ring Binders
Hole Punch	Scissors	White-Out

The cost savings of going paperless extends outside your walls to both sides of your global supply chain. Can you interact with your vendors more efficiently without paper? Are you still faxing invoices back and forth? Are you handing your customers a three-ring binder's worth of paper when they close on their homes instead of a USB memory stick with all the important data, including scanned images of the signed pages? That USB will fit nicely in a safety deposit box for easy retrieval when it comes time to sell the home. Some customers may want the three-ring binder full of paper. Give it to them. But don't give others what they don't want, or what they won't be able to find when they need it.

IMPROVING PRODUCTIVITY We can debate whether going paperless improves productivity (it does) or whether saving a few trees matters (collectively, it does). But going paperless absolutely will reduce costs. Whatever your budget is for office supplies, launching your paperless campaign will reduce it immediately by 25 percent. Empower the clerical staff to lead this effort. Start with not ordering new printer cartridges for desktop printers at $30 apiece. Later, you'll see money savings from fewer filing cabinets and perhaps even less need for more office space.

THE MYTH OF THE PAPERLESS OFFICE

In January 2007, a tremendously innovative video by Michael Wesch, "Web 2.0 . . . The Machine Is Us/ing Us," was released on YouTube and quickly became one of the most popular videos in the blogosphere.[34] It visually takes the viewer from paper, pencil, and eraser to digital text, which is obviously the key tool in the paperless revolution.

Also in January 2007, television commercials for two different U.S. companies show the opposite ends of the paper vs. paperless continuum. The first, for an office products company, shows two colleagues fulfilling their New Year's resolution to get their office organized by buying filing cabinets, bookshelves, three-ring notebooks, and shredders to deal with all the paper on their desks. The other commercial, for a printer/scanner company, has a group of colleagues in horror after learning that a sprinkler in the file room destroyed all their records, financial and otherwise. "How will we ever be able to comply with regulatory and audit requirements?" they moan. Fortunately, their IT colleague had previously scanned all the documents and stored them electronically for retrieval anywhere at any time. Unless the first company bought waterproof filing cabinets, score one for the IT guy.

The computer industry has been telling us for years that computers will enhance our productivity, and lots of statistics bear this out. Although computers, wireless, and the Internet may have simply created a false sense of productivity growth—perhaps we are producing more because we are working 24-7, not producing more in the same amount of time. What computers certainly have not done is make paper unnecessary. Walk into any large company's headquarters, see the rows of file cabinets, and you may think that the paperless office is perhaps less likely—if that is possible—than before the computer age. Computers make it easier to produce paper, and employees are doing just that. So this ease, coupled with new regulatory requirements on publicly traded companies (e.g., Sarbanes-Oxley), has yielded more paper and more filing cabinets, bookshelves, and three-ring notebooks to hold it all.

It doesn't have to be that way. Paperless doesn't mean the end of storage, just the end of the need for filing cabinets. Compliance is often much better

---

34 This video was retrieved May 12, 2008, from http://youtube.com/watch?v=6gmP4nkOEOE.

served with digital storage because the management oversight component of compliance works best when corporate data can be shared quickly and simultaneously to multiple people, rather than in a manila folder passed office to office with the hope that nothing falls out. Computers don't generate more paper—employees do. Explain the cost savings of a paperless office, reward desired behavior, and employees will produce less paper. Start by setting up a digital suggestion box for employees to offer ideas to reduce the amount of paper used in the office.

> **MAKING SMALL CHANGES** Just like books, newspapers, and magazines, paper in offices isn't going away anytime soon either. Focus on making small changes that collectively will change the culture in your company. Don't put the new phone list in everyone's mailbox—send it to them as an e-mail attachment. They can print it and put it up in their cubicle if they wish. Get rid of individual printers on employee desks—make printing a little harder and people will find alternatives. Clean out the file cabinets and give them away.

Equally anachronistic is the still all-too-common fax machine. Sascha Segan explains this staying power this way: "Think about it. Fax machines sell for $80 a pop. They don't get viruses or spyware. Their interface is a phone keypad, plus one (usable) button. When they don't work, the reasons are usually pretty obvious. And they feed stacks of paper."[35] In this regard, the fax machine is similar to the book. Ease of use has kept it from being replaced by some newer form of technology.

If you are not set on storing your paper in manila folders stuffed in file cabinets or storage boxes, your company may want to explore online storage options. My personal option—external hard drives sitting on my home office and office desks—works for me because I'm looking for a back-up solution only. If you want others in your company to be able to update or retrieve information to or from stored records, you have two paperless

---

35 Segan, Sascha. (2007). *Death to the Fax Machine.*

options. The first is a network drive. Just as your personal computer has a C: drive, so too can it have an E: drive that is not actually resident on the hard drive, but on a server somewhere. If this is your own server, you will need to offer your employees a way to access that drive remotely while not allowing unauthorized eyes to sneak a peak. This can be done very securely, but there are costs involved. A less secure solution is online storage. Google is rolling out its Gdrive service, which allows access to information from both personal computers and mobile phones. A host of lesser-known firms offer a similar service. But do you trust Google, or anyone else, with your company's proprietary data?

Regardless of the trust issues at play, the online storage industry as a whole is currently experiencing tremendous growth. According to a recent IDC report, revenue for this emerging market will reach $715 million by 2011, representing 33.3 percent compound annual growth between 2006 and 2011.[36] The trends toward greater reliance on digital and online information haven't swept everyone along with them, though. Oftentimes you cannot change your customers' views of things such as fax machines, but you can change your own behavior. I have a digital fax service for which I pay nothing. (I would be charged if I wanted a local telephone number.) When someone sends me a fax to that number, I get a paperless attachment in an e-mail. It takes no extra effort for the sender—he doesn't even need to know I won't get a document—and it's free and paperless for me.

In *Whatever Happened to the Paperless Office*, Matt Bradley points out that he saw signs of progress in the near future, but he qualified the progress: "Since the advent of advanced and reliable office-network systems, data storage has moved away from paper archives. The secretarial art of 'filing' is disappearing from job descriptions. Much of today's data may never leave its original digital format."[37] Fewer tasks for your secretary can lead eventually to fewer secretaries and further cost savings.

---

36 As Businesses Go Paperless, Owners Face New Threats, Decisions. (2008, March 27). *MarketWatch*.
37 Bradley, Matt. (2005). *Whatever Happened to the Paperless Office*.

A QUICK AND PAINLESS TRANSITION Your business doesn't have to wait Moses's forty years for a new generation of employees to lower costs. Paperless doesn't have to mean no paper for your business; start with a goal of 25 percent reduction immediately and see how quickly and painlessly that goal can be accomplished. The initial 25 percent reduction mandate will not necessarily be universally well received— the Luddites will complain in the break room. The young, the green, and the clerical staff will embrace it first. But soon everyone will find new and better ways of doing their old tasks. You will find the second 25 percent reduction (totally 50 percent less paper in your office) will actually be easier than the first and you won't have to issue another mandate—your employees will accomplish it without having to.

PROMOTE A GREENER WAY OF DOING BUSINESS

We must not, in trying to think about how we can make a big difference, ignore the small daily differences we can make which, over time, add up to big differences that we often cannot foresee.

—MARIAN WRIGHT EDELMAN

With the paperless revolution, like many other green initiatives, progress in the United States is often offset by increases in consumption elsewhere, particularly the rapidly developing economies of Brazil, China, India, Russia, Vietnam, and much of Eastern Europe. Emerging markets offer hope for new revenue opportunities for paper sellers in the near term because the people there are perhaps not as comfortable with e-mail as the developed world is. But it seems very likely they will follow the same path as developed countries in migrating toward a more paperless society. The way emerging countries have embraced wireless, for example, makes this a safe prediction. Couple this with the cost savings going paperless offers and emerging markets will have another reason to go paperless. As with other commodities like copper and cement, demand for paper in emerging markets means higher prices in the United States, thus providing a further incentive to reduce paper consumption here at home. At some point, rising gas prices reduce the numbers of miles we drive; this will be true for paper consumption too.

After rising steadily over years, worldwide paper consumption has flattened in this century. In the richest countries, consumption fell 6 percent between 2000 and 2005, from 531 to 502 pounds (241 to 228 kilograms) a person.[38] The data bolster the view that the ever-increasing volume of paper consumed may not increase forever. Most of us print far more at the office than we do at home, probably because we pay the cost of printing at home. If managers could change that mindset, companies could realize substantial savings. Scanning is a great way to convert paper to digital files, but the green benefits cannot be realized unless we use less paper up front.

Certainly even a fully paperless world won't solve all our environmental problems. We may kill fewer trees, but our mobile phones, external storage devices, and big-screen monitors all use more energy than a piece of paper in a manila folder. Newton's third law of motion states that "For every action there is an equal and opposite reaction." So too with paperless—it is a means to several green ends, but not a panacea. But starting down the path toward a paperless office sets a green tone for your office and other pro-environmental processes will develop in turn. And you get to enjoy the cost savings right away.

In South Korea, the government has been at the forefront of an effort there to promote consumer confidence in paperless economic transactions. The Framework Work Act on Electronic Commerce, passed in 2002 and amended in 2005, legislated simplification of e-commerce business processes and practices. Led by the Ministry of Commerce, Industry, and Energy, the government promoted digital documentation as a means to achieve cost savings in minimizing the production, storage, and distribution of paper documents. Simply put, a certified e-document, such as a sales

---

38   Fairfield, Hannah. (2008, February 10). Pushing Paper Out the Door. *The New York Times.*

contract between a manufacturer and distributor, was now as legally binding as a traditional paper document.

It's telling, however, that one of the most digital nations on Earth couldn't move beyond paper for commercial transactions before the government changed regulations and created the Certified e-Document Authority (CeDA). An example of how the process works can be seen in figure 4.1. In this illustration, the patient visits the hospital. Rather than handing him a prescription, the hospital sends that prescription to a secure server that in turns sends it to the pharmacy of the patient's choice. There is no chance for a criminal to change the prescription or forge an entirely false prescription. And it gives the pharmacy trust that the prescription is genuine. A similar e-prescription application could likely find a home in the United States, where drug-related fraud is much higher than in South Korea.

Figure 4.1 A Paperless Prescription Service in South Korea[39]

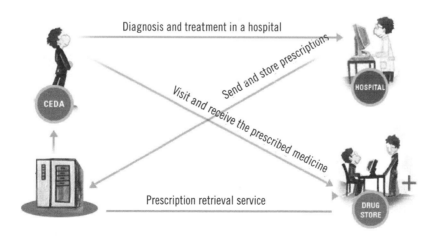

---

39   Retrieved on May 12, 2008, from http://www.ceda.or.kr/Eng/e04.jsp.

Whether your motivations for promoting a paperless office are to facilitate the flow of data, reduce costs, or help your company go green, I caution you that you'll have to be ever diligent or face the charge of hypocrisy from your less-enlightened colleagues.

THE PAPERLESS REVOLUTION has no downside to your company because people can still print what they want if they choose to. The upside is cost savings and being more environmentally friendly. Another clear benefit of going paperless is that the bottom of the organization can lead the effort since they are often the biggest producers of paper. Get them involved by asking what time-consuming paper-related task they would like to eliminate. How would they do it? How much time would that save you every quarter? How much money would it save the company? Then implement these ideas (even the more obvious ones), reward the employees who made the suggestions, and enjoy the cost savings and improved morale from empowering the less powerful in your organization.

If you would like to advocate the paperless position in your workplace, may I offer three tips (other than just practicing what you preach):

1. The first step must be to rid your space of the surfaces paper could easily occupy. Get rid of all filing cabinets and freestanding bookcases. No printer on your desk either. Paper will multiply if you allow it to.

2. Back up your hard drive regularly and store those back-ups away from your computer. Since you are not backing up your operating system (like Windows XP or Vista) or the core applications (like Microsoft Office), this is a quick and painless process. I have an 80 GB external hard drive in my office and at home and once a week I back up all my files to both. So all my files exist in three places and I can avoid those Luddites who anxiously await a hard-drive crash to remind me that their manila folders never crash.

3. Remember, paperless doesn't mean no paper. However, reducing the paper you use should remain an ongoing goal. Using recycled paper isn't enough, given how much energy is used to recycle it. But we must recognize that refusing to use any paper is simply not viable in our world today.

PAPERLESS WEB 2.0

Paper cannot extinguish a fire. —**CHINESE PROVERB**

One feature of Web 2.0 seems to be enhancing aspects of paperless in the form of digital libraries for users' content. For example, Flickr (www.flickr.com), the online photo-sharing site, is a successful example of a Web 2.0 application with paperless consequences. (Of course, *successful* doesn't necessarily mean *profitable* in Web 2.0-speak.) Let's say you are on vacation and using the hotel's Internet connection to stay connected. You can upload your holiday photos and your friends and family back home can see them. This is easier than trying to send the photos or video to everyone individually via e-mail. And it is certainly easier, faster, and less expensive than waiting until you go home, bringing your memory card to a local drug store, printing out multiple copies of the photos, and using snail mail to deliver them to your friends.

The migration of this Web 2.0 application (photo sharing) to a Paperless Web 2.0 application might be as simple as auto-uploading, meaning you take a picture with your camera phone and it will be automatically sent to a Flickr-like site. You will be able to tag that photo so that friends who might be interested will be notified on their mobile devices that there is a new photo available. There will be no need to use a traditional personal computer. This picture could be sent to a social networking site too, like MySpace or Facebook. And your friends could view it in real time on their mobile devices. The appendix includes a list of some of the current Paperless Web 2.0 offerings.

Web 2.0 is inherently digital and thus inherently paperless. Web logs (blogs), online diaries that encourage discussion and rebuttal, are already widely influential in the worlds of politics, communication, and celebrity.

But companies can employ them as a continuous, paperless perception survey, both internally with employees and externally with customers. Would you like to get your employees involved in a discussion about possible changes to your dental plan? A blog on your intranet can promote a candid—and, if necessary, anonymous—debate on the tradeoffs between cost and benefits. Wikis, which differ from blogs in that the original content can be changed rather than just commented on, are a paperless way to create better user manuals. Instead of writing such a manual and circulating it through departments for comments and edits, free wiki software lets everyone in your company try to improve the document. Really simple syndication, better known now as "RSS" and an aspect of nearly every blog or website, is another paperless tool to facilitate information sharing. This feature notifies interested parties if you add content to a blog or change a wiki.

DISTRIBUTE CONTENT DIGITALLY Web 2.0 is digital and thus intrinsically paperless. Keep it that way. Don't print your digital pictures and then snail mail them. Send them electronically or upload them to a photo-sharing site and let the recipient decide what format the pictures should be in. Distribute all content digitally. Recipients can convert those bits into atoms, or keep the content in a digital format to share it quickly and easily with others.

CONCLUSION

For a successful technology, reality must take precedence over public relations, for nature cannot be fooled. **—RICHARD FEYNMAN**

Dunder Mifflin, the fictional paper supply company featured in the U.S. television series *The Office,* is officially "as green as we have to be." On their website, www.dundermifflin.com, the company proclaims, "Dunder Mifflin is committed to improving the environment. That's why we plant a tree for each and every metric ton of paper that we ship. We look at it as an investment in the future because without trees, we have no paper and without paper, we have no business."

Just like the hardworking folks in *The Office*, I must confess that I'm also only as green as I have to be. I didn't go paperless to save a tree; I went paperless for the same reason both you and I send e-mail instead of writing letters: it is easier, faster, and cheaper. Since it's true for e-mail, might there be other things in your office that would be easier, faster, and cheaper if done without paper?

If your motivation for starting your company on the path toward paperless is to go green, that's wonderful. Your customers and employees don't wake up in the morning hoping you'll kill more trees than you have to. But if you don't care about green, go paperless anyway. Your data will be easier to store and retrieve, greatly enhancing your company's productivity. Both sides of your supply chain will also see gains in efficiency. You'll spend less money on dozens of office supplies. And yes, you may save a tree.

CHAPTER 5
PAPERLESS OUTSIDE THE OFFICE

The greatest danger for most of us is not that our aim is too high and we miss it, but that it is too low and we reach it.—**MICHELANGELO**

Technology shifts are either revolutionary or evolutionary. The automobile, the personal computer, the mobile phone, and the World Wide Web were all revolutionary, prompting seismic shifts in the way we live our lives. The iPod, the smart phone, desktop icons, and Bluetooth headsets were evolutionary—all nice to have, better than what was before, but none of them led to fundamental change to the planet. A revolutionary technology is a disruptive technology; we don't know that it is revolutionary until years after the early adopters started singing its praises. A successful evolutionary technology is greeted with "It's about time."

Paperless technology is evolutionary. The benefits of going paperless are obvious from the very beginning. Nonetheless, the evolutionary path toward paperless in many specialized industries will impact us much more so than the everyday paperless office. The use or rejection of paper in the publishing, education, government, law, medicine, and personal finance industries touches us every day, whether or not we spend any time in an office.

PUBLISHING

Outside of a dog, a book is man's best friend. Inside of a dog
it's too dark to read. —**GROUCHO MARX**

The obituary for the daily newspaper has been written again and again by Web 2.0 pundits. Some chuckle that all that remains is the decision of where to publish that obituary. It seems like just about everybody in both the entertainment and the technology worlds believes that all media will shed their analog past and thrive in a digital future. Newspapers were slow to become websites, but they are all online now, smothered with ads in an effort to offset the dramatic loss in revenue from print ads. Photos are becoming JPEGs. Songs are becoming MP3s. The question remains, what does this mean for the book and the magazine?

Nothing evokes a bigger reaction to the argument for going paperless than the notion that it will mean the end to books and magazines. I don't see a product on the horizon that I'd take to a sandy beach in lieu of a traditional book. I subscribe to several weekly and monthly magazines, even though the content is often available digitally at the same time. I enjoy tucking them under my arm to read during lunch or while waiting in line. And I get a lot of my news online now, but I still subscribe to the local paper. The paperless revolution will not end the availability of the written word on paper anytime soon. What it will mean is that some content available today on paper—phone books come to mind—will no longer be available, just as new music is not distributed on 8-track or cassette tapes and only rarely on vinyl, even though music purists swear they can hear the difference between analog and digital music.

Think instead of the paperless revolution just as you think of the revolution in digital music. When compact discs (CDs) were the monopoly format for digital music, prices soared, and consumers had to buy a whole CD when all they wanted was a song. The .mp3 digital compression standard and file-sharing sites like Napster ended that monopoly and consumers were able to obtain a single song. The tremendous success of iTunes shows that people will still pay for digital music. Note that iTunes songs that sell for $0.99 each generate roughly the same per-song revenue as a whole CD, but offer consumers the chance to choose what songs they want to buy.

The hardware piece of the equation has been the biggest stumbling block to broader acceptance of written digital content. Amazon.com, one of the world's largest sellers of traditional paper books, began selling its new e-reader, the Kindle, in November 2007. The Kindle is about the size of a large paperback book, but looks like a laptop computer. It has a special screen to increase the readability of the text. It is not a laptop—you can't create PowerPoint presentations on it. But you can surf the World Wide Web and download e-books.

> It's not hard to see how Kindle will take off. Business travelers will be the first to embrace it. Having a device with multiple books, newspapers, magazines, and blogs to travel with, which also has a long battery life, beats wrangling a laptop, magazines, and papers in an airline seat. The next market will be university students. With such a nifty application and the tension over ridiculously high prices for textbooks, going digital is a brainy way to deliver textbooks to an audience that is already used to digital consumption. . . . Books don't have to end and neither do authors' revenue stream. If I sell my Kindle book to a reader for $9.99, he has saved perhaps $20 on the price of a hardcover book.[40]

People like me who believe that single-function devices will be replaced by a Swiss Army knifelike device, can easily dispute the $20 savings per book claim. The Kindle costs about $400, so the first e-book you buy for it costs $410, not $10. This means that the first twenty e-books you buy for your Kindle actually cost $30 each ($400 for the Kindle, plus twenty times $10 for the e-books equals $600, divided by twenty equals $30 each). The first forty will cost $20 each. So by the time the hardware has been amortized over the purchases of enough e-books (say forty) such that the price per e-book is

---

40 Kiley, David. (2007). *Amazon Can Empty Bookstore Shelves.*

less than an actual book, you will be ready to replace it with a multifunction handheld mobile device.

E-reader hardware needs software, and in this case, the Kindle needs content. Following the iTunes model, Amazon.com now sells the Kindle and digital content for it on its site. It promotes the value proposition this way:[41]

- Revolutionary *electronic-paper* display provides a sharp, high-resolution screen that looks and reads like real paper.
- Simple to use: no computer, no cables, no syncing.
- Wireless connectivity enables you to shop the Kindle Store directly from your Kindle.
- Buy a book and it is auto-delivered wirelessly in less than one minute.
- More than 115,000 books available, including more than 100 of 112 current *New York Times*® Best Sellers, for just $9.99, unless marked otherwise.
- U.S. newspapers including *The New York Times, Wall Street Journal,* and *Washington Post;* magazines including *TIME, Atlantic Monthly,* and *Forbes* can all be auto-delivered wirelessly.
- More than 300 top blogs from the worlds of business, technology, sports, entertainment, and politics all updated wirelessly throughout the day.
- Lighter and thinner than a typical paperback.
- Holds over 200 titles.
- Long battery life. Leave wireless on and recharge approximately every other day. Turn wireless off and read for a week or more before recharging.
- Unlike Wi-Fi, Kindle utilizes a high-speed cellular data network, so you never have to locate a hotspot.
- No monthly wireless bills, service plans, or commitments.
- E-mail your Word documents and pictures (.jpg, .gif, .bmp) to Kindle for on-the-go viewing.

---

41   Retrieved May 12, 2008, from http://www.amazon.com/.

The Sony reader, which predates the Kindle by a year, uses the same six-inch *electronic paper* that the Kindle does. It too is trying to do for the book what the iPod did for music—enable "the paradise of portable digital consumption."[42] But in the "what were they thinking" category, you can't search the text of a book using the Sony reader. (Amazon's Kindle does have this function.) Why would consumers migrate to e-books if they fail to provide the most basic advantage of digital text?

THE PUBLISHING INDUSTRY While it might seem obvious that booksellers would start to offer e-books, contrast this with the tortuous path that music companies went down before offering online music. E-books won't eliminate the demand for traditional paper books, but they will clearly offer a new distribution channel with high margins. Once a book has been written, the biggest costs are printing and shipping it. The e-book product eliminates printing and shipping and thus can be offered either at a lower price or with better margins or both.

Whether or not e-readers become more commonplace, the paperless revolution will mean—because the content is digitized—consumers can pick and choose what they want. If I read only the sports section and am willing to read it online, why should I have to pay for the whole paper and the cost to deliver it? If I want a single chapter of a text for my students, why can't I have them buy just that chapter without the time, trouble, and expense of photocopying that chapter and binding it to others? The incremental revenue to the publisher is actually higher because I'd never ask the students to buy a whole book just to read a single chapter, but I might ask them to buy one chapter from ten different books.

In Japan, where long commutes on public transportation are the norm and there exists a vastly superior mobile phone network compared to what we have in the United States, new services are constantly being rolled out:

---

42 Grossman, Lev. (2007). *Reading Gets Wired.*

Sales of mobile-phone novels—books that you download and read, usually in installments, on the screen of your mobile phone—have jumped from nothing five years ago to over ¥10 billion ($82 million) a year in 2007 and are still growing fast. Mica Naitoh, a popular author whose bestselling book had 160,000 downloads a day, says many of her readers never even buy old-fashioned books. For one thing, she says, today's trendy handbags are far too small.[43]

There are cultural factors in Japan—those long commutes and a near 100 percent penetration rate for mobile phones—that make reading novels on a mobile phone more likely than in other countries, but obviously a market exists, at least when a dedicated e-reader device is not required. In the United States, where the mobile penetration rate is smaller than in Japan and the use of public transportation is substantially less, the Kindle might find a market. But I doubt it.

Without enough time passed to measure the Kindle's success, the trend in the United States still seems to be using either a laptop or a mobile device. Hearst Corp.'s *Seattle Post-Intelligencer* plans to test software developed with Microsoft to let readers download an entire newspaper onto laptops and mobile devices running Windows software. Once the content is received, readers can view the material without being connected to the Web. Hearst said it may use the software to provide downloads of its other newspapers and magazines.[44] Somehow I don't think techie news junkies want old news offline.

The bottom line is that books aren't going anywhere anytime soon. Consumers like them. A book does seem to be an almost perfect product, at least from a technology standpoint. It turns on instantly, doesn't need a battery, and can be used by even the youngest consumers without an instruction manual.

Not everyone agrees. Bill Hill, Microsoft's point man on e-reading, offers this:

We chop down trees, transport them to plants, mash them into pulp, move the pulp to another factory to press into sheets, ship the sheets to a plant to put dirty marks on them, then cut the sheets and bind

---

43 *Japan's Latest Mobile Craze: Novels Delivered to Your Handset.* (2007).
44 Lazaroff, Leon. (2007). *Software Lets Readers Download Newspapers.*

them and ship the thing around the world. Do you *really* believe that we'll be doing that in fifty years?[45]

I appreciate the irony that as the author of *The Future of Less*, I am siding confidently with those who think the future of the printed book is assured. For a new product to replace an existing one, it must do everything the old one did and some things the old one did not. The car replaced the horse and buggy because it could still get you from here to there, but did it much faster. There is a place for e-readers and e-books, just like you can still ride a horse. But there were more cars sold last year than there were horses ridden.

EDUCATION

Education is what remains after one has forgotten what one has learned in school. —**ALBERT EINSTEIN**

The current generation of middle and high school students, like my daughter and son, are generally more tech savvy than their teachers. I shudder when I see my kids tasked to buy an atlas, trace the continents five times, and label all the countries. The labeling I understand for a geography class, but let them find and print blank maps from the Internet.

Well beyond having to buy an atlas, almost daily I see my kids asked to do things the way their teachers did them when their teachers were in school. Certainly this is slowing the migration to a more paperless classroom. More importantly, it is a signal that some teachers are refusing to adopt technology to teach a new generation of learners that as a group would almost certainly be responsive to digital learning. Teachers have long feared that technology will replace them, something that is not going to happen anytime soon. We're social beings and learning certain kinds of things is simply better when we're taught by a human. But what will eventually happen is that teachers who can use technology will replace those who cannot.

---

45 Ibid., p. 64.

THE PAPERLESS REVOLUTION IN EDUCATION Public education faces tightening budgets and thus the cost savings that going paperless affords is critical. But for-profit entities can learn something from this model. The paperless revolution in education is also about changing the way you do business to address the needs of a new customer base. My children receive assignments via e-mail when they miss class. Paperless technology likewise allows you to talk to your customers when they are not able to receive your traditional brochures.

I teach MBA students, and now all of my classes are completely paperless: no textbook, no hardcopy papers to turn in, and students can take notes directly on their laptops, which the school provides to all students as part of tuition. As an interim step before abandoning the textbook, I experimented with Zinio (www.zinio.com), which allows the author to protect his intellectual property through a basic DRM scheme while giving the student the feel of turning pages, being able to highlight relevant passages, etc. Some of the big textbook publishers now offer a similar product. The biggest advantage is that the textbook could be downloaded for about 40 percent of the price of the hardcopy version; the disadvantage is that students can't resell the book at the end of the term. In the end, my students—hardly a scientific sample—said they hated reading the e-textbook on their laptops and printed it out anyway. A better e-reader (e.g., Amazon.com's Kindle) may partially alleviate that issue, but I've already moved on,

For a course I teach on managing technology, there is a test on about fifty vocabulary words and terms that I think managers need to know, like *disruptive technology*, *802.11*, and *WiMAX*. I gave the students a traditional hardcopy test where they filled in the blanks and short-answer segments in ink. For extra credit, I asked them a question I didn't know the answer to:

Given what we have discussed about wireless technology, peer-to-peer communications, Bluetooth, etc., please explain how I could have offered this exam paperlessly, but still prevented students from using their computers to look up the answer or help other students. This is a technology—not an ethics—question.

None of their answers completely satisfied all the ways that electronic cheating could occur. The most common answers—don't allow students to connect to the Internet via a cable and turn off the wireless capability that exists in all our classrooms—doesn't solve other student-to-student connectivity opportunities like peer-to-peer 802.11, Bluetooth, and cellular text messaging.

In 2007, the Massachusetts Institute of Technology (MIT) announced it would make available the university's entire 1,800-course curriculum by year's end. Currently, some 1.5 million online independent learners log on to the MIT site every month (www.ocw.mit.edu/index.html) and more than 120 universities around the world have inaugurated their own sites for independent learners. MIT has more than 1,500 course curricula available online to date.[46] This is certainly wonderful news for those who want information available to more people, more easily. This kind of sharing is not practical—and barely possible—without the digitization of information. But since what was formerly written content—a syllabus, reading list, class notes, and PowerPoint presentations—is only one piece of an education, I'm certain enrollment at MIT has not declined. The opportunity for one-on-one, face-to-face interaction with faculty eliminates the threat to MIT and others.

For those with a different vision from MIT's, the Internet is still forcing the old guard to go paperless. It used to be that a team of researchers would submit their results to a journal. A journal editor would then remove the authors' names from the paper and send it to their peers for review. Copyright rested with the journal publisher. Not anymore. The Internet—and pressure from funding agencies who are questioning why commercial publishers are making money from government-funded research by restricting access to it—is making free access to scientific results more and more of a reality. Though the academic community didn't choose to do so on its own, it has been forced to accept that much of the value of knowledge depends upon the wide distribution that only the World Wide Web can offer.

Virtually every top-fifty MBA program, like the one where I have the good fortune to teach, uses the case study method. Scholars go into the business community to research real-world issues and problems and write cases for students to try to come up with their own solutions. In North America and for English-speaking programs in general, there are two primary sources

---

46   Gardner, W. David. (2007). *MIT to Put Its Entire Curriculum Online Free of Charge.*

that sell cases, the Harvard Business School (www.hbsp.harvard.edu) and Canada's Ivey School of Business (www.cases.ivey.uwo.ca). Harvard allows professors to choose the cases they want to use for a course and bundle them together online. Students then pay Harvard directly (cases are usually $3–7 each) and then download them. DRM protection is employed. In contrast, Ivey prints and mails each case to students, slowing down the delivery and accomplishing nothing in terms of protecting the content. Students could scan the paper and share the cases anyway—we are teaching the Napster generation. My guess is that Ivey will be forced by its competition to migrate to paperless soon too. It would make a good case study.

What is happening in education matters to you because these are your future employees and customers. The habits they develop in college will impact how they interact with their bosses and how companies will have to interact with them if they want to earn their business.

GOVERNMENT

Without computers, the government would be unable to function at the level of effectiveness and efficiency that we have come to expect. This is because the primary function of the government is—and here I am quoting directly from the U.S. Constitution—"to spew out paper."

—DAVE BARRY

Aside from defense and public safety, U.S. federal and state government agencies are notoriously slow to adopt new technology. Generally, they seem to wait until the commercial marketplace has chosen its winners and prices have fallen as manufacturers reach economies of scale either through mass production (think personal computers) or through the commoditization of standards-based products (think 802.11 access points). As a taxpayer, I don't mind this trend to wait until things get less expensive. Nonetheless, we are making progress toward a more digital and paperless government.

The U.S. federal government has been trying to use the Internet to reduce paper, which has—not surprisingly—resulted in cost savings and increased efficiency. In the February 2002 budget submission to Congress, the office of the U.S. president outlined a management agenda for expanding e-government:

> E-government does not mean putting scores of government forms on the Internet. It is about using technology to its fullest to provide services and information that is centered on citizen groups. We know that the public expects this kind of service from the government, and that it uses the Internet more than ever before. Polling data from the Pew Foundation, for example, show that over 40 million Americans went on-line to look at federal, state and local government policies, and over 20 million used the Internet to send their views to governments about those policies.[47]

Almost six years later, some cost savings from going paperless are finally being realized in Washington. The White House announced that it is going paperless when it submits the fiscal 2009 budget. Instead of printing 3,000 free copies of the budget for the media, lawmakers, and the White House and Cabinet, the 2,200-page document will go online at www.budget.gov. The change was announced via e-mail. "This step will save nearly twenty tons of paper, or roughly four hundred eighty trees," the White House press secretary said. "In terms of fiscal savings, we estimate the e-budget will save nearly $1 million over the next five years."[48]

PAPERLESS GOVERNMENT With the transition to paperless business, the U.S. government can have fewer physical offices for people to visit, or at least fewer employees working there. The explosive growth in filing taxes online shows that we will go paperless if you make it better than the alternative. Are there processes in both ends of your global supply chain that could make it easier for your customers to be your customers?

---

47 Retrieved May 12, 2008, from http://www.whitehouse.gov/omb/egov/g-1-background.html.
48 *Paperless Budget Will Save Cash, Trees.* (2008).

According to a February 2007 review of progress to date, the government reported selected examples of how citizens and federal employees are benefiting through the use of these e-government initiatives:

- GovBenefits.gov provides a single point of access for citizens to locate information and determine potential eligibility for government benefits and services. It receives approximately three hundred thousand visits per month and to date has provided nearly 1 million citizen referrals to benefit programs.

- IRS Free File, in the 2006 filing season, was used by more than 3.9 million citizens to file their taxes online for free. But almost 80 million returns—about 50 percent of the total—were filed electronically using commercial programs. IRS electronic filing began in 1986.

- Grants.gov has seen a substantial increase in the number of grant application packages posted. In the first quarter of fiscal year 2005, 252 packages were posted on Grants.gov. By the last quarter of fiscal 2006, 4,523 packages were posted.

- SAFECOM increased the number of urban areas that can establish interoperable communications at the command level within one hour of a major event from ten to seventy-five. Additionally, 66 percent of public safety agencies report using interoperability to some degree in their operations.

- The U.S. Department of Labor, one of the first agencies to complete its migration to an e-travel service provider, reported a savings of $35 (almost a 60 percent savings) per trip by creating paperless travel orders for employees, and a reduction in processing time from seven to three business days. Additionally, the Departments of Transportation and the Interior are each realizing savings of over $1 million per year through the use of online booking.

- More than eight hundred forty thousand federal employees have registered in the GoLearn.gov learning management system. Federal employees have completed more than 2.7 million courses provided through the e-training initiative. E-training is assisting the federal government in maintaining a highly skilled workforce at a fraction of the cost of solely classroom training.

- Through migration to fewer payroll service providers, agencies are able to realize cost savings and improved efficiencies. For example, the U.S. Department of Health and Human Services has reduced the annual costs of payroll processing for its more than sixty-five thousand employees from $259 to $90 per employee for an annual savings of almost $11 million.

- Each month, more than one hundred thousand resumes are created on USAJobs.gov, the federal online recruitment service. USAJobs. gov receives more than two hundred forty visits daily from job seekers looking for information regarding career opportunities with the Federal government.

In January 2008, the U.S. government announced it would provide prepaid debit cards to Social Security recipients, which should save money for both parties. The officials said card users will have faster access to their money and will be protected by additional security features. "We've been working for a while to try to understand the needs of the unbanked," said Judith Tillman, commissioner of the Treasury's Financial Management Service, which distributes most payments from the government. "Combine that with problems we've seen with financial crimes and identity theft, problems with forged checks and stolen checks and so on—the debit card seemed like the right answer."[49]

Florida, where I live, has greatly reduced the amount of paperwork required to live here. I can renew my driver's license online, pay the fees for my license plates, and purchase my annual business license all without filling out one piece of paper. The state does charge me $7.50 to mail me a paper copy of my business license, which I wouldn't need to pay if I could download it instead. More and more services are moving online, which means less waiting in line. I have no complaints about that.

And it is not just about money. Official government records, paperless digital records, stored remotely would have made some aspects of the recovery from, for example, hurricane Katrina more efficient. The U.S. government has offices all over the world. On behalf of the states, they can store birth and marriage records in the event that local records are destroyed. This is how much of corporate America backs up its files.

---

49 *U.S. Plans Social Security Debit Cards.* (2008).

LAW

I use the rules to frustrate the law. But I didn't set up the ground rules.

—F. LEE BAILEY

One of the universal images of a law office is bookcases lined with rows and rows of legal volumes that seem designed to impress or intimidate those outside the profession. Certainly those books were outdated the moment they were printed, as new legal precedents are being set almost every day. Less impressive but much more efficient would be an online system that was accessible anywhere, anytime. It would be less expensive to produce, more current in its information, and an obvious tree saver. But, "It's hard to imagine a career less conducive to conserving paper than that of a lawyer."[50]

The cliché that "ignorance is no excuse under the law" has certainly slowed the migration to paperless legal documents. Consumers simply don't know what requires a traditional ink signature on a traditional piece of paper and when an e-mail or e-fax will suffice.[51] According to Ethna Piazza, a lawyer and partner at Sheppard, Mullin, Richter & Hampton LLP, e-signatures of various sorts are perfectly legal. "Every time you click 'buy' on an online shopping site, for example, you're executing an e-signature." And although e-mail isn't the best medium for executing contracts, it's possible to agree officially to something over e-mail.[52] But in a profession like law with a low-risk tolerance, change is slow.

> The feeling of informality in e-mail communication has lured more than one communicator into an embarrassing or legally risky error. I suspect that the first Egyptian Pharaoh who put pen to paper immediately sensed the permanency provided by such communication. However, I don't think the Pharaoh had the problem of hordes of hungry lawyers seeking discovery of his scrolls.[53]

---

50 Perkowski, Mateusz. (2006). *Practically Paperless Lawyer.*
51 15 U.S.C.A § 7001 (http://www.law.cornell.edu/uscode/uscode15/usc_sec_15_00007001-
   ---000-.html) states that "a signature, contract, or other record relating to such
   transaction may not be denied legal effect, validity, or enforceability solely because it is
   in electronic form." You should consult an attorney for additional information.
52 Segan, Sascha. (2007). *Death to the Fax Machine.*
53 *Death by E-Mail.* (2007).

A lawyer dictating notes to a secretary to type up later is a recurring image in many television shows and movies. But this too is changing, albeit—again—slowly.

Digital dictation is a win-win for the lawyer and the law firm. The lawyer still generates billable hours while dictating, but can now do so outside of the office, on the way to another opportunity for additional billable hours. And since most lawyers can talk faster than their secretaries can type, they save time for themselves and their assistants. The law firm is a winner because it can hire fewer secretaries or put their time to more productive uses.

As a group, lawyers have not been early adopters of new technology. The BlackBerry seems to have a caught on as a way to increase billable hours, but two words best define how receptive the legal profession generally is regarding technology: Word Perfect. Lawyers were among the last profession to abandon that software and probably did so only because their clients had long ago abandoned it. But the need for cost savings—as in other professions—continues to drive change.

> **THE LEGAL PROFESSION** Law is a profession with ancient rituals, language, and practices. Paperless will not come soon to this industry. What is a meme in your company or industry that is long overdue for change? Have you ever asked yourself, "Why are still we doing it the same way my father did?"

MEDICINE

All sorts of computer errors are now turning up. You'd be surprised to know the number of doctors who claim they are treating pregnant men.

—ISAAC ASIMOV

As the son of a pharmacist, I witnessed firsthand the "Can You Read This" quizzes in my father's professional magazines, meant to test a pharmacist's ability to decipher a doctor's penmanship on the prescription blank. The obvious potential risk in misreading what was written seems to me to be a great motivation for the medical profession to migrate to paperless. Digital prescriptions—sent directly from the doctor to the pharmacist of the patient's choosing—would also reduce fraud and could mitigate some risk to non-English-speaking patients. A recent medical drama on television featured a Mexican woman who overdosed on her medication because she mistook *once daily* for eleven pills daily, *once* being "eleven" in Spanish. According to the CEO of Allscripts, a digital electronic health records company, "Prescription errors injure 1.5 million and kill 7,000 patients annually and most mistakes could be avoided if scripts were written electronically."[54]

Even within hospitals, handwritten script writing poses risk to patients. The logical solution, called computerized practitioner order-entry (CPOE), is, according to the *Journal of American Medicine*, "widely regarded as the technical solution to medication ordering errors, the largest identified source of preventable hospital medical error." Studies show that adopting this system can decrease errors by up to 81 percent.[55]

Nonetheless, despite years of private-sector pressure on U.S. hospitals to install CPOE systems, the number using such systems remains a small fraction of nonfederal hospitals, according to the latest counts by two key IT market watchers.[56]

One incentive for medical professionals to migrate toward a paperless environment is the 1996 Health Insurance Portability and Privacy Act (HIPPA). This act mandates the adoption of "standards for transactions, and data elements for such transactions, to enable health information to be exchanged electronically," and regulates the use and disclosure of confidential information about patients' health.[57] One fear was that digitized records might be too easy for hospitals to share with, for example, insurance companies.

54 Kher, Unmesh. (2007). *Chasing Paper from Medicine.*
55 Koppel, Ross, et. al. (2005). *Role of Computerized Physician Order Entry Systems in Facilitating Medication Errors.*
56 Retrieved May 12, 2008, from http://www.hipaadvisory.com/action/eHealth/index.htm#cpoe.
57 Retrieved May 12, 2008, from http://aspe.hhs.gov/admnsimp/pl104191.htm#1173.

THE MEDICAL PROFESSION Medicine is highly regulated and issues of compliance—both to the government and to insurance companies—mean lots of forms to fill out and patient records to maintain. Going paperless is better because that information can be shared more quickly when a patient is out of town or in the event of an emergency. Going paperless can likewise help your business prepare for emergencies by making content digital and accessible remotely.

According to Gartner Group research, the problem with the U.S. healthcare industry as a whole is how little it spends on information technology. This is despite patient deaths following hurricane Katrina that were attributable to lost paper medical records. For anyone who has visited a doctor's office or the hospital recently, it doesn't take a lot of imagination to see that health information technology can improve the efficiency, cost-effectiveness, quality, and safety of medical care.

Getting less attention, but headed in the right direction, is dentistry. My dentist now uses digital X-rays rather than film and has computer terminals in every room to view them. But I still get a postcard instead of an e-mail to remind me to show up for my next cleaning.

It is not just about the digitization of content to make information easier to share. When the focus is on the medical community speed equals the chance to save a life. For your business, the speed that going paperless affords may mean the chance to win a new customer or keep an existing one.

PERSONAL FINANCE

Many people want the government to protect the consumer. A much more urgent problem is to protect the consumer from the government.

—MILTON FRIEDMAN

Checks are among the lone paper survivors in the personal finance market, but they too will soon be gone. Consumer use of checks peaked somewhere in about the mid-1990s, according to the U.S. Federal Reserve, but remained the dominant means of noncash retail payments until 2003.[58] The migration from checks to credit and debit cards, direct deposit, and automated clearing-houses has greatly reduced the cost associated with processing each transaction. The further migration to a cashless society will reduce costs further and reduce fraud.

Attempted check fraud has more than doubled since 2003, to $12.2 billion in 2006, according to a study by the American Bankers Association (ABA). Actual losses were up 43 percent to $969 million. And advances in laser printing are spurring a rise in counterfeit checks, which now account for 28 percent of losses. The ABA and the U.S. postal service have launched FakeChecks.org to educate people about such ploys.[59] Going paperless would eliminate all of this.

Identity theft experts have been telling us for years that sending sensitive financial documents through the mail is a bad idea. There is risk of theft at both ends. Paper account statements and credit card bills give mail thieves all the information they need to conduct a fraudulent transaction and damage your reputation as well as your bank account. You can lessen the danger by buying a locking mailbox and depositing outgoing mail at the post office. But going paperless is a better way to enhance security. Online bill pay is safe and secure since the information I send to my bank and the information I receive from them is encrypted while in transmission. Should someone intercept my information en route, it would be nothing more than ones and zeros.

In addition to lowering your risk of identity theft, there are many other good reasons to consider making the move to paperless personal finances:[60]

- Save on stamps.
- Save time. When you get a paper bill, you have to get your checkbook, write a check, put it in the envelope with the stub, put a stamp on it, and walk out to your mailbox.

58 *The Use of Checks and Other Non-Cash Instruments in the United States.* (2002). *Federal Reserve Studies Confirm Electronic Payments Exceed Check Payments for the First Time.* (2004).
59 Der Hovanesian, Mara. (2007, December 17). *Check that Check.* p. 18.
60 Derived from *Five Ways Paperless Personal Finance Saves Your Money.* (2007).

- Save on balance float. Paperless banking allows you to schedule payment at the last minute, keeping those funds in your account for as long as possible.

- Save on late fees through e-mail notification of balance due and the fact that electronic payments are rarely lost.

If you are interested in migrating your personal finances from your mailbox to your e-mailbox, here is a useful five-step plan:[61]

- Reclaim your mailbox.

- Use OptOutPrescreen.com (https://www.optoutprescreen.com/) to stop credit card and insurance offers.

- Stem the tide of junk mail with the Direct Marketing Association (https://www.dmachoice.org/MPS/mps_consumer_description.php).

- Cancel unwanted magazine subscriptions.

- Consolidate accounts and close those you no longer use. Reduce the number of credit cards you carry. If you have bank accounts at multiple locations, combine them at a single bank. The fewer accounts you have to track, the less paper you have to deal with.

- Use electronic billing. If you have a choice between paper and paperless, opt for the latter. Not only will this reduce clutter, but it can also save you money from some firms.

- Computerize your checkbook. This is probably normal for young adults, but many people still balance their checkbook by hand.

- Scan all documents and receipts and shred the originals. Most scanners automatically convert documents to .pdf files. If you know which financial records to keep and how long to keep them, you can purge (shred) your archives.

My own experience with paperless personal finances has been a good one. I charge everything I can to a single credit card to earn frequent flyer miles on my preferred airline. Instead of getting paper bills from my service providers—wired phone; mobile phone; DSL; home alarm; life, home, and car insurance; and a handful of others—I receive all these bills via e-mail or

---

61 Derived from *In Pursuit of Paperless Personal Finance*. (2007).

via an e-mail notification to read them on a secure website. I review my credit card charges online every month and pay directly from my bank, online. For someone like me who travels frequently, the ability to manage my finances when out of town is a particularly useful advantage to being paperless.

CONCLUSION

The illiterate of the 21st century will not be those who cannot read and write, but those who cannot learn, unlearn, and relearn. —**ALVIN TOFFLER**

Inside or outside the office, the advantages of being paperless are pretty much the same—the faster exchange of information, lower costs, and better security. Inside the office, however, you are either setting or following policy and working with your colleagues to improve processes. There is nothing wrong with that. But outside your office, you have much more control. I do business only with companies that offer paperless billing and online bill paying. If your company wants my credit card, car/life/home insurance, home security, limousine, pest control, or pool cleaning business, you must be paperless. Consumers can drive reluctant companies to adopt paperless policies that will actually save those companies money.

I chose to discuss the vertical markets above—publishing, education, government, law, medicine, and personal finance—because all are illustrative of industries where paperless has gained a foothold and is growing. What's next? What, years from now, will we look back on and say, "Do you remember when we used to do this with paper?" An April 2007 announcement from China made clear to me how even small changes to the way we do things will make us ask this question. China Southern Airlines, in conjunction with China Mobile (the world's largest MNO, with about 400 million subscribers), launched a mobile e-boarding pass, a service where your boarding pass is sent to your mobile phone. A scanner near the check-in counter reads the bar code in the text message and the passenger gets a paper boarding pass. After a trial period, it seems likely that the paper pass can be eliminated for intracountry flights.

Paperless doesn't mean no paper. In this example, it just means no boarding pass. I predict that soon paperless will mean less paper in virtually every industry. The attitudes of new employees and new customers—a generation that grew up with digital content and seems more aware of what is happening to the planet—will make this transition essential because of the eco-friendly aspect.

SECTION 3

THE FUTURE OF CASHLESS

CHAPTER 6
LIFE WITHOUT CASH

I was a bank teller. That was a great job. I was bringing home
$450,000 a week. —JOEL LINDLEY

nvestigative reporter Edison Carter, the human component of the virtual *Max Headroom* in the 1987 television show of the same name, carried around a device the size of a pen that was used for all financial transactions. Insert the pen into a reader and credits were removed from his bank account. The tagline for the show was *20 minutes into the future*. That future—a cashless society—is available today, and the rapid growth of new wireless telecommunications technologies will hasten the eventual, ubiquitous migration to a cashless society.

In his bestselling book *Being Digital,* Nicholas Negroponte uses the first chapter to differentiate between atoms and bits. The difference between atoms and bits is a good place to start to understand the implications of a cashless world. Paper money and coins are atoms; they have to be carried, exchanged, sometimes repatriated, and eventually replaced with new paper and new coins. Digital money is also carried (probably in an electronic wallet or some other electronic device), exchanged, and sometimes repatriated, but it never wears out, gets torn or lost, and—with good security—can never be stolen.

Security and the lower cost to mint digital money would seem to encourage people to clamor for its hasty implementation. Although a December 2007 Google search for *cashless society* yielded 133,000 hits, many of them equated it to facilitating tax increases (it would at the very least be a lot harder to avoid paying your taxes) or hastening the apocalypse because of the mark of the beast system to be installed (the mark being a microchip or something very similar). However, as science-fiction author Philip K. Dick said, "Reality is that which, when you stop believing in it, doesn't go away." Whether or not everything on the Internet about a cashless society is true, there does not seem to be much public pressure to get policy makers to support ending our reliance on cash. The U.S. penny may go away soon, and I'll bet anyone two bits that more change will occur.

Why are we not further down the path toward a cashless economy, given the available technology? *The Economist* offers a partial answer: "The challenge, as so many pioneers have found to their cost, is to predict, and then promote, a change in human behavior. Typically, such changes are triggered by accident."[62]

The answer is not a lack of need: The world didn't need the iPod, the SUV, or the bacon double cheeseburger, but they have all found broad acceptance regardless.

The answer may be a combination of generational and social factors. In the United States, government policy makers tend to be older and not as receptive to new technology. Their mindset seems to be "If it ain't broke don't fix it." Socially, Americans have been slow to both demand and receive the latest technology offerings, compared to Asian cultures. It appears that it might not be just the U.S. policy makers who want to leave well enough alone. The case needs to be made, in the simple words of David Warwick, that "the world is hardly getting along fine with cash."[63] Cash engenders crime. Cash transactions often hide criminal activity. It can be counterfeited. It might promote disease. And cash doesn't work everywhere unless you exchange it for the local version.

---

62 *Dreams of a Cashless Society.* (2001).
63 Warwick, David. (2004). *Toward a Cashless Society.*

CREDIT CARDS—
THE FIRST CASHLESS HARDWARE

Technology feeds on itself. Technology makes more technology possible.

—ALVIN TOFFLER

In *Looking Backward* (1888), Edward Bellamy wrote that a cashless society would exist at the end of the twentieth century. That vision is what exists today in the form of a traditional credit card. Almost one hundred years later, George Morrow predicted the following:

> As almost all of us routinely use credit cards to make purchases, we have already made a radical departure from our traditional cash-oriented society. A logical extension of this trend should bring us a credit card–size, dedicated computer that performs all personal financial transactions. Looking ahead, this card—and it can't be far off—will identify you, give you and the bank your personal audit trial, balance your checkbook when you plug it into a phone line, buy food, clothing and houses for you. It will pay your rent and your utility bills. You will no longer have any need at all for cash—ever.[64]

As a cashless society became technically possible, Reynolds Griffith speculated in 1994 on practical pros and cons.[65]

ADVANTAGES

- Theft of cash would become impossible. Bank robberies and attacks on shopkeepers, taxi drivers, and cashiers would simply cease to occur.
- A change to recorded electronic money would be accompanied by a flow of previously unpaid income tax revenues, allowing rates or the national debt to be lowered.

---

64 Morrow, George. (1984). *A Computerized Cashless Society.*
65 Griffith, Reynolds. (1994). *Cashless Society or Digital Cash?*

DISADVANTAGES

- Privacy—companies would have the ability to gather data on buying habits.

- Counterfeiting would seem to be much easier with such electronic bank notes than with paper money, since they are merely electronic signals.

- There is not (in 1994) a mechanism to easily permit transfers from one person to another.

Fifteen years ago there was not a good mechanism to easily transfer digital currency from one person to the other. Now there is: our Swiss Army knifelike mobile phone makes for a great digital wallet. Counterfeiting would be much harder, not easier, given advancements in color photocopying since Griffith made his predictions. Electronic bank notes will have more security than paper notes ever had. Adding smart chips to credit cards or other physical devices is an interim step to improving the security of cashless banking until cards themselves are replaced with a more secure mobile device. What hasn't changed is the privacy that using cash offers.

In December 2007, the U.S. Federal Reserve undertook a study to measure trends in noncash payments in the United States. It determined that the number of noncash payments was over 93 billion in 2006, up from less than 82 billion transactions in 2003. And electronic payments exceeded two-thirds of those transactions, with traditional checks making up the balance.[66] The reported concluded, "Electronic payments are being used more frequently in transactions where checks or cash may have been used in the past."[67] It is likely your own habits mirror this. Are you writing fewer checks now than you did five years ago?

Long before I started thinking about the cashless revolution, I remember a set of actions I saw happening at grocery stores and finding it odd. People would withdraw money from an ATM located in the store and hand it to the cashier, who placed it in a drawer. Then the manager took that cash and deposited it in a bank, which took the money to refill the ATM. I was sure that this labor-intensive Rube Goldberg operation could not last forever.

---

66 *Non-Cash Payment Trends in the United States 2003–2006.* (2007). p. 4.
67 Ibid., pp. 4-5.

Old habits die hard. There were 5.8 billion ATM withdrawals in the United States in 2006 with a value of $578 billion. The average withdrawal was almost $100.[68] Why would anyone use cash—with all the associated risks (e.g., loss, theft)—when seemingly better alternatives exist?

The answer is probably more cultural than technological. Prior to the explosive growth in debit cards, cash was a way to avoid going into debt. It forced one to live within one's means. Many Americans are now heavily in debt and perhaps that explains the more recent growth in debit card usage compared to credit cards. It is possible that some people are losing interest in accumulating more debt, or perhaps have at least lost access to more credit. So the debit card is effectively paperless cash and a return to the pay-as-you-go philosophy.

> **CREDIT AND DEBIT CARDS** The best things about credit and debit cards are that they are easy to get, easy to carry, and easy to use. They don't need an instruction manual or your kids to explain them to you. Whatever replaces credit and debit cards will have to do all this and more. More will mean integration into a handheld device with biometric security with an interface that can buy a newspaper out of a box on the street, can be used to pay your friend the ten bucks you owe him from losing a bet on the ball game, and being able to increase your limit or transfer money to your account while sitting at a restaurant with your biggest potential new customer.

---

68 Op Cit., *Non-cash Payment Trends in the United States 2003–2006.* (2007). p. 13.

Debit cards have now overtaken credit cards in terms of the number of transactions, but still represent less than half the dollar value of credit card transactions (see figure 6.1).

Figure 6.1 Number and Value of Noncash Payments in 2006[69]

	Number	Value
Checks	33%	55%
Debit Cards	27%	1%
Credit Cards	23%	3%
Automated Clearing House (ACH)	16%	41%
Electronic Bank Transfer (EBT)	1%	0%

Since wireless technology is the foundation for successful paperless and cashless revolutions, it is not that surprising that South Korean consumers are at the forefront of the cashless revolution. According to the Korean government, in 2006 South Korean consumers spent about $250 billion on their credit cards, nearly half of all private consumption. And although South Korea ranked just 34th in per capita income among countries in 2005, it ranked fifth in per capita credit card spending, according to the Bank of Korea, the country's central bank.[70]

Following the Asian economic crisis that hit South Korea in the late 1990s, a newly elected government started a campaign to fight the corruption and tax avoidance that cash transactions make possible. As part of its effort to fight the free flow of cash in the underground economy, the government encouraged consumers to use credit cards and threatened tax audits of enterprises that refused to accept them. One consequence was that South Korea became (and remains) one of the world's most credit card–friendly countries. So when the government mandated that stores accept credit cards, South Korea's tech-savvy, entrepreneurial spirit kicked in to make paying electronically as easy as paying in cash. In 2005, there were about 400,000 electronic cash registers per 1 million people, compared to fewer than 11,000 in Japan. The advances in these electronic cash registers and their ubiquity enabled

69   Ibid., p. 9.
70   Choe, Sang-Hun. (2007). *To Save, Koreans Use Credit Cards*

South Koreans to pay for virtually anything with credit cards: parking tickets, highway tolls, pizza deliveries, even a 2,000-won bill (about $2.20) at a street-corner noodle shop.[71]

CARDLESS CREDIT

What has always been will not necessarily always be. **—ROBERT GOIZUETA**

Credit and debit cards remain the dominant cashless consumer devices, but there is a growing number of other devices that accomplish the same function as a credit card in a different form.

The Speedpass key fob, for instance, allows you to buy gasoline without a physical credit card. The consumer simply holds the device near the receiver, and the credit card the user used when registering the device is charged with the transaction. It can be used inside the gas station, too, for nonfuel purchases. Speedpass promotes its product by noting that "(The) Speedpass key tag has a built-in chip and radio frequency antenna that allows it to communicate with Speedpass readers . . . (It) is safe and secure. Your card information, preferences, and personal details are not stored in your Speedpass device, so your information is protected from unauthorized use."[72]

The goals of simplicity and convenience are met and will be enhanced as the Speedpass morphs into one more function on your mobile device and thus can be used virtually anywhere, like we see in Japan and South Korea. Security is more problematic. A lost fob can be reported and deactivated, but this is less desirable than real-time, pre-expenditure security based on some type of biometrics, the science of using biological properties to identify individuals (e.g., fingerprints, retinal scan, and voice recognition). To enhance security, Speedpass added the requirement to enter one's zip code ("zip code verification") to complete a transaction. The Speedpass is still faster than a credit card because most cards now require similar zip code verification. This minor security tweak is an improvement, but it is still possible for the bad guys to beat the system.

---

71 Ibid.
72 Retrieved May 12, 2008, from https://www.speedpass.com/forms/frmHowItWorks.aspx?pgHeader=how.

Another cashless innovation for speed is the toll tag, an electronic device in a vehicle that allows drivers to pass through tollbooths without stopping. Introduced in the 1980s in the United States, its popularity is rising in urban areas worldwide. Because the user is often required to maintain a prepaid balance, they can sometimes get a discount when using it instead of paying with cash at the tollbooth. Many cities worldwide offer special lanes for tag users so that they don't have to slow down or queue behind nonusers.

While you might not think of the toll tag as one of the greatest inventions of the last twenty-five years, its ability to allow you to avoid the cash lines at tollbooths is a perfect model for other markets. The grocery store has a ten-items-or-fewer lane—how about a no-cash lane to really speed things up? A similar device could speed you through subway turnstiles. And earlier we discussed getting on a plane without a traditional boarding pass—you'd get to the overhead bins first.

The advance of a cashless economy isn't limited to speed-oriented credit card replacements, of course. Consider Tokyo, which is famous for, among other things, ubiquitous vending machines. Unlike many of the biggest cities in the United States, the vending machines in Japan are almost always outside and accessible 24/7, most likely due to the unlikelihood of vandalism or theft in that culture and the fact that many Japanese work and play late into the early morning, when most stores have closed.

CARDLESS CREDIT Today, cardless credit is relegated to a handful of single-function devices like the Speedpass, a toll tag, or a fob that fits on your key chain in lieu of a credit card. But just as we've seen standalone digital cameras, GPS, and MP3 players morph in a multifunction mobile phone, these standalone cardless credit devices will be integrated into some kind of multifunction device, almost certainly a mobile phone. And instead of just working in a single venue, they will be accepted anywhere that a credit or debit card is today. This means no more cash advances to your employees, no more lost travelers' checks, and no more corporate credit card accounts to monitor and manage.

All the way back in 2001, Coca-Cola, the Japanese wireless operating giant NTT DoCoMo, and Itochu Corporation launched "Cmode"—information terminal vending machines in Japan that allow customers to purchase a Coke using only their mobile device. The machines allow customers to accumulate user points that can be exchanged for soft drinks or ring tone downloads. And in a possible hint at a way to use government funding to offset some of the costs of nationwide deployment, the machines are equipped with a speaker that would allow them to disseminate information such as disaster updates from local government agencies.[73]

There are advances from nations outside Asia, as well. In early 2003, France launched computerized smart cards targeted at smaller transactions that would otherwise require pocket change. Because the basic Moneo card is anonymous, there are no privacy or identity theft concerns. But if an owner loses his or her smart card, the cash that's stored onboard can be used by whoever finds it, which is why there's about a $100 storage limit.[74] Something similar might prove to be the answer to the privacy issues associated with going cashless.

THE FUTURE OF E-COMMERCE

The only way of discovering the limits of the possible is to venture a little past them. —**ARTHUR C. CLARKE**

Money on the World Wide Web has always been a tricky and—due to the huge amount of commerce that takes place at a distance and anonymously—a hugely important consideration. It has also been searching for alternatives to cash almost from the start. Although some early sites, like eBay, allowed users to send checks or money orders for purchases, this has all but disappeared in favor of secure, cashless payment systems. And it is not surprising that some of the largest e-commerce sites have their own payment systems. This is likely to change as news reports of data theft from online merchants stress out U.S. consumers who, not long ago, ordered products online and then phoned in their credit card information.

73   *The Unwired Coke Machine.* (2001).
74   *Cashless Society Gets Mixed Reviews.* (n.d.).

The process now is bad and is certainly the product of the rapid growth and independence of the early days of e-commerce. Today, most consumers have sent their credit card information to every merchant they do business with, just like they are forced to hand their card to every merchant they do business with in the non-online world. But e-commerce should (arguably, must) be better than the offline world in terms of security because of the lack of a real relationship between buyer and seller.

> **ONLINE TRANSACTIONS** Only a few years ago, people would order something online and then call the 1-800 number to tell some stranger their credit card number because they thought that phone call was somehow more secure than entering their number online. Now we are paying our taxes online with every personal financial detail about our life being delivered to the U.S. government. This whole Internet thing? It could be huge. Talk to everyone in your organization about how they use the Internet to shop or otherwise interact with merchants. What best practices can you adopt from what others are doing? What annoys you the most about e-commerce? Make sure your site is not similarly annoying your soon-to-be ex-customer.

Fortunately, e-commerce is getting better. Google and industry leader PayPal (owned by the giant auction site eBay) represent a step in the right direction. Instead of sending your credit information to everyone, with Google Checkout or PayPal, you store it one time with your trusted service. That service pays your e-commerce transactions and debits your card without the actual vendor ever seeing your card number. So you have to trust one company instead of trusting everyone you are doing business with. If you don't have this level of trust, you probably aren't shopping online anyway.

It is possible that such services can evolve to also perform the kind of ID checking required at traditional brick-and-mortar merchants to purchase such items as alcohol and cigarettes. Perhaps future e-commerce could verify identity and age: you send Google a copy of your passport and they certify your account is for someone over twenty-one years old. With this kind of assurance, e-commerce sites could move beyond asking customers to "certify" that they are of age and begin to actually enforce age-restriction laws. It's possible. Already the government has delegated, by necessity, its oversight role of the sale of alcohol to the checkout clerk at the corner 7-Eleven. Why not have Google and PayPal replace the driver's-license checkers?

MOBILE CREDIT

Technology is anything that was invented after you were born. —**ALAN KAY**

What form will an increasingly cashless society take, though? The mobile phone is already the dominant consumer device facilitating the migration to a cashless society in Asia (most notably in Hong Kong, Japan, Singapore, and South Korea) and Europe. The consulting firm Frost & Sullivan estimates that the Asia-Pacific region will account for more than a third of the global mobile commerce market by 2009 and Europe for more than 27 percent.[75] You can already spend the whole day in Tokyo, for example, without carrying cash or credit or debit cards and instead paying for everything, including consumer goods, with just your mobile phone. Worldwide payments using mobile phones will climb from $3.2 billion in 2003 to more than $37 billion by 2008.[76] The largest cities in South Korea are equally advanced. The trend is clear.

This rapid growth is easy to explain when you consider that there are many advantages to all three parties involved in mobile commerce—the users, the banks, and the MNOs.[77]

- For users, mobile commerce facilitates and reduces the cost of sending and receiving money through a bank and its associated fees, while still enabling financial transactions without the risks associated with the use of cash (e.g., theft).
- For banks, mobile commerce provides them with an opportunity to further enhance their customer reach by moving the unbanked

75 Amin, Shaker. (2007). *M-Banking—To Bank the Unbanked.*
76 *A Cash Call.* (2007). p. 71.
77 Amin, Shaker. (2007).

community toward banked status. Some people will never have enough money to open a bank account or don't live close enough to a bank to make it worthwhile to be a customer. But a mobile phone doubling as a savings account means the bank has a customer it never would have otherwise.

- For MNOs, with their advantageous position as the first customer point of contact, mobile commerce allows them to offer their growing subscriber base new services. Since the core competence of MNOs lies in delivering mobility solutions, they will have to partner with financial institutions in order to gain access to credit facilities, credit payment management, and other financial services. If you wanted an iPhone on day one in America, you had to be an AT&T customer. Similarly exclusive marketing arrangements could be made between MNOs and banks.

In Asian countries, embracing the cashless revolution has meant using mobile networks to deliver services to subscribers. There's a pattern in the way each nation accepts these changes, with reactions and progress clustered in four distinct groups:[78]

- **The leaders:** Japan and South Korea. E-cash is replacing paper money.
- **The mobile tigers:** Hong Kong, Singapore, and Taiwan. Mobile commerce is in limited use, mainly for transportation services. Poor use of excellent wireless networks to support mobile payments (m-payments), simply payments from wireless devices.
- **The giants:** China, India, Indonesia, and the Philippines. These countries must deploy mobile commerce at a slower rate due to their massive populations and far-flung geography. Mobile commerce is in use for remittances, bill payment, and mobile ticketing.
- **The midmarkets:** Thailand, Malaysia, and Vietnam. Bigger in population than the global mean and with better than average wireless networks, these countries will be close followers of the giants.

Japan is among the world leaders in going cashless. In early 2007, the largest MNO in Japan, NTT DoCoMo, introduced *osaifu-ketai* (portable

---

78 *Mobile Payments in Asia Pacific.* (2007). p. 4.

wallet) functionality to its customers, which allows mobile phones to function as an ATM card to withdraw cash, a credit card with charges simply added to the existing monthly bill, train pass, plane ticket, and other items.[79] Starting from next to nothing in 2006, twenty million subscribers in Japan now have e-wallet functionality in their mobile phones. It's important for the MNO to continue building out its partner network because the financial success for DoCoMo is not in the phone sales, but in the per-transaction fees.[80] In other words, if I use my phone to buy a Coke and am charged the equivalent of 50 cents on my bill, the MNO pays Coke only about 90 percent of that, keeping the rest as profit.

CONVENIENCE AT THE CHECKOUT Paperless doesn't mean no paper, but cashless does mean no cash is your business ready to go totally cashless or to deal with customers and vendors who are? Even if you don't care, your customers might. Don't you hate it when you're at a business that doesn't accept your credit card? *What kind of golf course doesn't take Amex,* I asked recently at a local clubhouse. No big deal, but I've never been back. Are you doing business with your customers the way they want to do business with you? Existing and new entrants in the online payments systems, like PayPal and Google Checkout, seem to be providing an important service—what lessons can your business learn from how they grew their businesses? Japan and South Korea are seeing explosive growth in the number of cashless commercial transactions—how can your business position itself to use the knowledge of what is going on over there to change the way you interact with your customers and suppliers? How will the rapid growth of mobile banking and mobile e-commerce impact your business?

The market for cashless applications in Japan is huge. Per capita credit card usage there is just one-seventh that of the United States, but stored value (i.e., prepaid accounts) is much more popular. As of March 2007, mobile

---

79 Parmelee, Nathan. (2007). *DoCoMo's Holding Strong.*
80 Ibid.

wallet penetration was already approximately 40 percent of NTT DoCoMo's fifty-two million wireless subscribers.[81] One explanation for the popularity of cash replacements in Japan is that the lowest paper-money denomination is one thousand yen, or about $8.80, making coins more common and somewhat less convenient for low-value payments compared to the United States and its ubiquitous one-dollar bill. Stored-value mobile wallets will eventually become popular in the United States once merchant acceptance grows, especially in the youth and other underserved segments with less access to traditional bank cards. Evidence for this exists in the proliferations of often exploitive payday lender business—working-class Americans often don't have ready access to many bank services. But, according to NetBanker, it likely won't reach current levels of Japanese penetration until 2013 or beyond.[82]

> What appeals to the Japanese about e-cash is the way it speeds things up. It offers the convenience and portability of cash, but more so. It takes no more than a tenth of a second to complete most transactions. As no change is required, counting errors are eliminated. Fraud and theft are reduced. For the retailer, it reduces the cost of handling money. And because e-cash is smart, it is easy to add extra services. For example, ANA, Japan's second largest airline, allows users to convert frequent-flyer miles into e-cash.[83]

Even though about half of all South Koreans have used their mobile phone to pay for products or services, the bulk of consumer spending is still done with credit cards. But it isn't much of a stretch to predict that, as with so many other things in South Korea, consumers will make the transition from physical to digital quickly and move from credit cards to an all-in-one mobile e-wallet. We can see this happening already in South Korea's huge mobile-gaming industry.

Korea's robust wireless networks all but created mobile gaming. For many years, Koreans have been the world's most prolific players of the tongue-twisting category massively multiplayer online games, using the country's excellent wired broadband network. It is not surprising that this application has been a huge hit

---

81   Bruene, Jim. (2007). *Mobile Payments Metrics: NTT DoCoMo.*
82   Ibid.
83   *A Cash Call*, p. 72.

with wireless subscribers too. And Korean MNOs are the beneficiaries of their subscribers using minutes to play online games.

Young people without a credit card have been targeted because they need a cashless way to purchase goods such as downloadable music, video, and attributes used in online gaming. Online merchants have taken this system and enabled users to make larger purchases as monthly limits have been increased from $20 to $120. As usage has grown, the system has expanded out to further enable users to pay for cable TV bills, newspaper subscriptions, and membership fees for clubs and associations. More than $1 billion is now charged directly to phone bills instead of traditional credit cards in Korea.[84]

Similar to what we see in Japan, the biggest piece of revenue for the Korean MNO comes from the transaction fees associated with e-wallets.

One of the biggest challenges facing any new technology is that the beneficiaries of the current technology do all they can to prevent the growth of the new, disruptive technology. So who are the would-be corporate losers in a cashless society? Governments are probably not too concerned about getting out of the money-printing business because their macroeconomic monetary and fiscal roles will remain. Banks and credit card companies will likely flourish in the new system as the demand for their services increases with the freer flow of commercial transactions.[85]

The following are already available in Asia and will be available in the United States in the next couple of years:[86]

- Transportation companies will offer "touch and pay" access to ticket counters where a stored-value card is either attached to the mobile device or embedded in the SIM.

- Retailers will offer loyalty cards using a similar means of payment so as to reduce the amount of cash they have to handle and the risks associated with cash.

- Advertisers will build Web links into posters in trains and buses and on buildings that can be activated by mobile devices passing by them at close range to push them to advertising's third screen, the mobile device.

---

84 *Mobile Payments in Asia Pacific.* (2007). p. 11.

85 Michal Lev-Ram argues that MNOs are the bottleneck because they haven't figured out how to make money on the transaction. See: Your Cell Phone = Your Wallet. (2006). *Business 2.0.*

86 *Mobile Payments in Asia Pacific.* (2007). p. 5.

- Vending machine operators will sell soft drinks and other consumables by enabling payment by mobile device.

- Content providers, including music and information sites, auction sites, and Web 2.0 social networking sites, will become available anywhere, anytime to mobile subscribers.

Mobile network operators are obviously beneficiaries of the migration towards m-payments and m-wallets. In late 2007 in the United States, Verizon Wireless—one of the largest MNOs in the United States—announced a new mobile-banking service. It offers customers the ability to check bank account balances, transfer funds between accounts within the same bank, and review and pay bills for those bills already set up through the financial institution's online account service. It is a secure service with PIN and device lock-out capabilities in addition to the security that is already incorporated by the financial institution. And customers will need to register with their financial institution in order to use mobile banking.[87]

CONCLUSION

Chaos is the score upon which reality is written. —**HENRY MILLER**

Unlike paperless, which doesn't mean no paper, cashless really does mean no cash—no paper, no coins, no wooden nickels. Singapore is already there. While in Singapore researching this book in March 2008, I met the managing director of a large company. Despite his prestigious job, he had no money in his wallet and said that is often the case. In Singapore, everything goes on plastic. Singapore is small, with a 100 percent urbanization rate and an educated, sophisticated population. No wonder it made it to cashless first. Japan and South Korea aren't far behind. The technology exists in both places to go completely cashless (certainly in the urban centers) but some cultural issues—such as the occasional need for privacy—prevent a total conversion. These concerns for privacy and security are real and transnational. But the largest cities in Europe and the United States will be cashless, though likely not cardless, in my lifetime.

---

87 *Verizon Wireless Launches Mobile Banking Services.* (2008).

When coupled with powerful mobile devices, cashless transactions can be safe, easy, and convenient for the user. In chapter 1, I asked you what you carried when you left your house. My guess was that your answer was a wallet or purse, keys, and your mobile phone. I'm not guessing when I predict that answer will change in the next ten years for many if not most of you reading this book. Pick up your mobile phone in the morning, and out the door you'll go.

CHAPTER 7
SECURITY, PRIVACY, AND EASE OF USE

The government's been in bed with the entire telecommunications industry since the forties. They've infected everything. They get into your bank statements, computer files, e-mail, listen to your phone calls. Every wire, every airwave. The more technology used, the easier it is for them to keep tabs on you. It's a brave new world out there.

—GENE HACKMAN AS BRILL (*ENEMY OF THE STATE*)

In *Toward a Cashless Society*, David Warwick wrote that he believes only governments can create a cashless society: "True cashlessness will come about only if a government undertakes the project since only the government can put an end to the production and circulation of cash, and only the government can realistically administer an electronic replacement for cash."[88]

Warwick is certainly right about the role of government in the production and circulation of cash, but likely incorrect about the order of implementation. No government will ban cash without the overwhelming support of its

---

88 *Toward a Cashless Society.*

citizenry. Political reasons drive the reluctance, but a practical reason governs it: citizens who didn't support the mandate would simply continue to use cash as the medium of exchange. Much more likely is that consumer use of electronic cash will reach a level where printing new money is simply no longer necessary, just as most governments today no longer make gold coins. *The Economist* observed that the end of physical cash would occur when a panhandler could take donations on his mobile phone.[89] It would probably mean the end of wishing wells too, though coins will remain in people's possession forever, long after they can't be used in vending machines. Las Vegas already has electronic slot machines that are coinless. Hard hit will be congressmen who want to hide $90,000 in their freezer in Louisiana, as one famously did after receiving a bribe from an FBI informant in 2006. But he got reelected anyway and is still in office. Cashless may come late to Louisiana. Most likely, the transition to a cashless economy will not be revolutionary, but evolutionary, to the point where we simply stop using cash.

Ultimately, the success or failure of the migration to a cashless society will most likely be tied to three critical issues relating to the pace of consumer acceptance: security, privacy, and ease of use. Security must be bidirectional— users must feel their money can't be stolen, and vendors must be confident that this is a bona fide transaction. Privacy means that consumers will not be afraid that the vendor (or worse, some unseen corporate or governmental Big Brother) is monitoring their transactions any more than they already do through a credit or debit card. And ease of use simply means that consumers must prefer cashless transactions and the broader cashless lifestyle to cash.

SECURITY

Security is mostly a superstition. It does not exist in nature. Avoiding danger is no safer in the long run than outright exposure. —**HELEN KELLER**

Security and technology should be happily married, but the relationship has often been a stormy one when the consumer in-laws enter the house. The migration to e-commerce was slow due to consumers' initial reluctance to use credit cards online, despite the fact that these same customers would gladly

89 *Dreams of a Cashless Society.*

hand over their physical cards to a waitress who disappeared into a dark corner of a restaurant. Encryption and secure Internet protocols (e.g., secure sockets layer), as well as a more established online community, have largely overcome consumers' concerns, though there is still the occasional horror story of credit card numbers stolen from or by an online retailer.

Security in a cashless society must have all the strengths that the current credit card system has, plus other technology that gives consumers and vendors even greater confidence. The only improvement to credit card security until recently was the option to add one's picture to the front of the card. Biometrics to verify identity through fingerprints, retina scan, or voice recognition at the point of transaction would seem to be a step—arguably the critical step—in the right direction and is already being introduced. There are a handful of companies rolling out new products. Some plug into a laptop, for example, and the owner's fingerprint is required to boot up instead of just a password. Banks will certainly replace four-digit Pin codes with something that can't be beaten out of crime victims who take their card and head to the nearest ATM. Some early market research says that Americans prefer fingerprint scans to retinal scans and thus new products will likely head in that direction.

SECURITY IS PARAMOUNT A new product won't replace an existing product if it is only as good as the incumbent. The new product must be better—often, much better. Of the three factors impacting the migration to a cashless society—security, privacy, and ease of use—security is the most critical. We are talking about our money and we won't do anything unless we are confident that we are in complete control of the transaction. A credit card, phase I in the path toward cashless, is better than cash: if you lose it, it can be replaced; if it is stolen, you have limited liability for its misuse; and it can be used in almost every store in every country in the world. So the next phase in the path toward cashless will be just replacing the credit card with something we carry anyway: the mobile phone. What will make this better, and thus likely to replace the incumbent, is that biometrics on mobile phones will make them more secure than the cards we use today.

Traditional cards have a magnetic strip that can be read by criminals using a simple handheld device. There were several credit card number theft rings apprehended in the United States in the 1990s who were using this technology. Smart cards, in contrast, have an embedded microprocessor or a memory chip to store the owner's information more securely than the old magnetic strips and are much more popular in Europe than in the United States. The first large-scale adoption of smart cards was for French pay phones in 1983.[90] The European health insurance and banking industries use smart cards extensively. Every German citizen has a smart card for health insurance.[91]

> Even though smart cards have been around in their modern form for at least a decade, they are just starting to take off in the United States, where magnetic stripe technology remains in wide use. However, the data on the stripe can easily be read, written, deleted or changed with off-the-shelf equipment. Therefore, the stripe is really not the best place to store sensitive information. To protect the consumer, businesses have invested in extensive online mainframe-based computer networks for verification and processing. In Europe, such an infrastructure did not develop since the card carries the intelligence.[92]

At the very least, all the credit and debit cards in the United States need to become smart cards. Many experts are predicting that 2008–2009 will see the mass migrations from magnetic stripe cards to smart cards with consumers getting the new cards when they renew their existing cards, with American Express being an early leader. According to JupiterResearch, microprocessor chip cards are likely to be in more than 75 million American wallets before the end of 2008 and more than 128 million will be in circulation by 2009.[93]

Alternatively, the United States could leapfrog Europe and skip this whole generation of technology by using mobile phones instead. A mobile phone is the ideal electronic wallet because it already is:

- Unique to a user via its electronic identification number (not to be confused with a phone number, which is not hard coded into the phone);

90 *Will 2008 Be the Year of the Smart Card?* (2008).
91 *What Is a Smart Card?* (n.d.).
92 Ibid.
93 *Will 2008 Be the Year of the Smart Card?* (2008).

- Built with a wireless interface that could be used for authentication, authorization, and accounting purposes;

- Gladly carried by over a billion consumers;

- Customizable with the latest biometric security features; and

- Tied to a credit card account or some form of monthly payment, or prepayment in the case of young people or those in developing countries.

A mobile phone with biometrics could even be programmed to act as age certification for the purchase of controlled items like alcohol and cigarettes. One big reason for the delay in biometric mobile phones reaching U.S. consumers sooner is that we buy our phones from MNOs, who sell phones with features they want us to have—usually those that require us to use our billable minutes. In Europe and Asia, consumers can buy phones directly from the manufacturer.

Any real form of secure data must be founded on encryption. According to Allen Weinberg, managing partner at Glenbrook Partners, a financial services and electronic payments consulting firm, "It's relatively easy to make mobile phones very secure devices. Encryption is as good as or better than what you do with an ATM or at-home banking. No one is going to pick up your phone and start moving money around the world. It's just not going to happen."[94] Simply put, devices such as the Speedpass key fob go away (one less thing to carry) and the functionality migrates to the mobile phone with much better security.

PRIVACY

When Monopoly, the popular board game, replaced its play money with phony debit cards last summer, it was a sign of the times. But cash still boasts an advantage over plastic. It leaves no trace.[95] —*THE ECONOMIST*

Privacy remains the biggest stumbling block to a cashless society and the key to its success or failure. An entirely cashless society would require that sinners as well as saints adopt the technology. Criminals love cash. Not many folks buy

---

94 Fost, Dan. (2007). *One More Thing Cell Phones Could Do: Replace Wallets.*
95 *Cashiered.* (2007). p. 73.

illegal drugs with their credit cards. The government would certainly benefit from being able to track the illicit drug deals that are no doubt now conducted in cash, because cash gives both parties anonymity. So, too, does the anonymity of cash serve the buyer and seller of knockoff Rolex watches on the street corner, the babysitter who hopes to avoid paying taxes for her evening of labor, and perhaps even the largely decriminalized office football pool.

Certainly, there are also noncriminal transactions where one party would nonetheless prefer not to have an electronic record kept, such as buying an adult magazine, a gift for your girlfriend, or a bottle of scotch. David Warwick wrote that the migration to a cashless society will have to overcome consumers' need for occasional anonymity:

The march toward cashlessness stalled due to wariness about invasions of privacy. Scholars and public leaders alike are reluctant to consider the idea of replacing cash with an electronic currency system, particularly one operated by government, lest they be seen as willing to compromise privacy.[96]

As with security, people are probably willing to accept the same limit on privacy that they enjoy on a credit card, which is in fact quite little. Should the government want to, they can see where you purchased what and when, if you pay with a conventional credit or debit card. Prepaid debit cards, like the Moneo card discussed earlier, are probably the functional answer to this concern. Like today's prepaid debit cards, the level of anonymity approaches cash if the cards could be refilled without recording who the user is.

Again, the mobile phone offers a physical vehicle that is much less expensive than producing single-function debit cards. But the phone will always be perceived—and correctly so—as directly tied to the owner and thus not private at all. The mobile phone meets the "As Good as a Credit Card" test, but no better. My guess is that in countries where privacy can drive consumer behavior, like in the United States, prepaid stored-value cards will maintain their niche (and thus may one day be sold at a premium to face value!) for those who insist in a cashless world that they maintain as much anonymity as they had with cash. It might be a little harder for my babysitter to avoid declaring her income. I may have to pay her a little more. That doesn't bother me.

---

96 *Toward a Cashless Society.*

ANONYMITY To feel the impact of the potential loss of privacy from a truly cashless society, let's choose a hypothetical at the extreme end of the spectrum, however unlikely it may be. Let's say you are a prominent elected official who wants to hire a prostitute and take her on a business trip out of town with you. Paying her in cash eliminates the paper trail. If there were no cash, how could this or—more importantly—a lawful, but private transaction (e.g., buying a bottle of scotch or a carton of cigarettes) occur? This is a problem that can be solved.

Another option, of course, is some kind of digital representation of cash. The early Internet flirted with several variations of e-cash when consumers were still reluctant to enter credit card information online. The overwhelming majority of e-cash products offered by various companies, however, are not private electronic payments and not as anonymous as cash. When a user wants to purchase an item from a merchant, he sends the bank a special electronic message encoded with a unique digital signature requesting the money. The bank debits the user's account and sends e-cash to the merchant. After receiving the e-cash, the user's computer transmits it to the merchant's computer, which verifies the authenticity of the e-cash with the bank and credits it to the merchant's account.[97] So the bank knows which merchants its customers are doing business with—again, as good as a credit card, but no better.

Since the government arrests the drug dealers, wants to collect taxes from my babysitter, prints the money, and passes laws about who can buy what products when, it remains a critical player in the path toward a cashless economy. Privacy is never mentioned in the U.S. Constitution, but is firmly ingrained in the psyche of Americans.

> The move to a cashless society stretches the parameters of current legislation and legal precedent regarding the right of financial privacy. Congress must enact a new federal statute to balance the competing interests raised by digital cash. A viable digital cash statute is necessary

---

97 Downey, Catherine M. (1996). *The High Price of a Cashless Society: Exchanging Privacy Rights for Digital Cash.*

for users to fully embrace this new currency, and the statute must place the protection of individual privacy as its highest priority. Without such legislation, the Internet may never realize its full potential.[98]

Though the Constitution may afford us no guarantee of privacy, our political and economic system does empower the individual. If the government isn't passing laws we like then we have the chance to elect a new government. If merchants are not respecting our privacy, they'll lose our business.

EASE OF USE

Making a system easier to use for someone does not, for me, make that system better. You bring a "user experience" to life by designing with people, not for them. Users create knowledge, but only if we let them.

—JOHN THACKARAY

Home Depot now features several self-checkout lanes per store, where the consumer scans items and pays with cash or credit. The advantage to the store is that one employee can operate the equivalent of four registers simultaneously, while still making sure the customers scan all the items they place in their bags.

Imagine, however, taking this process and making the entire transaction cashless, reducing theft, and further reducing the number of employees needed at checkout. As a customer enters the store, the radio frequency identification (RFID) tag on his credit or debit card—or mobile phone—is read by the checkout system authorizing that customer. The customer places all the items he wants to buy in his shopping cart and walks out the door to his car. As he exits the store, the RFID checkout scans the items, charges the credit card, prints a receipt (or e-mails it to the customer), and rewrites the tags to show that those items were paid for and when to allow returns, warranty claims, and other time-related issues.

Giant retailers like Wal-Mart have been pushing RFID as a way to improve global supply management, including real-time in-store inventory counts. This could benefit consumers too if, for example, they could go online, access

---

98 Ibid.

that real-time inventory database, and see if a specific item was in stock. Better still would be the ability to use your mobile phone to scan a bag of apples and see when and where they were picked. Or scan a shirt to see the country of origin for the fabric and the country of manufacture. Customers could scan toys to see if they had been recently inspected to ensure no lead paint was found. Or they could scan a product to find out its recent sale prices (i.e., the last time it went on sale). RFID tags, unlike bar codes, can store more and more current information, can be updated regularly, and can be focused on providing information that consumers insist on.

The widespread deployment of RFID tags in everything sold to anyone seems trapped for now in the chicken-or-the-egg dilemma. Retailers says they need RFID tags in everything to make it worth their while to buy RFID readers, train employees, and in general change their business practices. But retailers want the cost to go down to less than a penny per tag. Companies who make RFID tags say they can get to these price levels if manufacturers committed to putting them into everything. But manufacturers want to see the prices fall before trying to pass that cost on to retailers.

The fast-food industry, which refers to itself as *quick service restaurants* (QSR), is all about speed. Its cash registers are designed so young people with limited job experience can process customers quickly, and those with poor math skills don't have to calculate how much change to give back. But that doesn't solve the problem of slow customers. In 2004, McDonald's announced an agreement with MasterCard to accept the credit card company's PayPass card, a contactless payment option that uses RFID to allow users to tap their cards on a reader, rather than swiping it and signing the receipt. For transactions of less than $25, no signature is required. This is where the biggest time savings comes in, especially with drive-through customers.

"McDonald's is always looking for new and innovative ways to use technology to improve customer service in our restaurants," said the McDonald's vice president of information technology. "The convenience of MasterCard PayPass will help our customers get their food even faster." . . . "In today's fast-paced world, people count on the convenience and utility of electronic payment options," said a MasterCard spokesman.[99]

---

99 *McDonald's Expands Cashless Payment Options with MasterCard PayPass.* (2004).

Since that launch in 2004, cashless transactions are now very common in the QSR space, not to mention 7-Eleven and other convenience stores, movie theaters, and other venues that place a premium on processing people quickly to reduce the time they stand in line. The use of contactless payments grew more than 15 percent in 2007, and the market is now valued at more than $200 million, according to ABI. It's expected to reach more than $820 million by 2013.[100]

> **MARKETPLACE CHANGES** A new product won't replace an existing product if it is significantly harder to use. Windows Vista has seen a lukewarm reception in the marketplace because the opportunity cost of updating from Windows XP seems too high—Vista might be better, but learning a new operating system seems hard. Credit cards were easy to use and quickly captured market share from cash. (They also allowed us to spend money we didn't have—another huge plus for many consumers.) Replacing credit cards with mobile phones will make credit or debit transactions easier and available in more places than even credit cards are today, such as vending machines, toll roads, and to get on the subway.

In the long-term, consumers are almost always the ones who determine how technology is implemented. If it doesn't work easily, they won't use it. "But what is worse for retailers is that the consumer may choose to go somewhere else, where ease of use was taken into consideration," said John Perry, a leading consultant in retail tracking and consumer use technology for retail.[101]

Again, this ease of use could come at a loss of privacy, with retailers knowing a lot about your buying habits. According to Nick Wreden, "The ultimate payoff may come in tracking customer purchases and establishing relationships through couponing and loyalty programs. This was a strategy [that] Exxon Mobil came to recognize early on in consumers' use of the

100 Malykhina, Elena. *Deployments of Contactless Payment Systems Slower Than Expected.* (2008).
101 *McDonald's Expands Cashless Payment Options with MasterCard PayPass.* (2004).

Speedpass technology."[102] But unlike Speedpass and existing frequent shopper cards, which monitor your purchases in exchange for a few rewards, it would be almost impossible to opt out of an RFID economy. That is the privacy issue that will almost certainly bother Americans the most. We can choose to surrender some privacy by, for example, buying our bottle of scotch with a credit card. But when someone else chooses to compel us to surrender some privacy—or worse, compromises our privacy without our knowledge—then Americans are going to push back.

Another step down the path toward a cashless economy is the rapid growth of gift cards—sometimes called *stored-value cards*—available at virtually every retailer and restaurant in the United States. Their ease of use—for both the giver and receiver—certainly contributes to the success they have found in the marketplace. On the surface, cash would seem to be better than a gift card since it could be used at any merchant. The old joke is that cash always fits and is never the wrong color. But it seems like both givers and receivers like gift cards. According to a 2007 study by Comdata Stored Value Solutions, the average gift card buyer will spend $203 on cards, a $17 increase from 2006. Teens have embraced gift cards faster than any other segment—98 percent of teens between fourteen and nineteen years old have either bought or received a gift card.[103] Don't look for gift-card growth to slow down anytime soon.

Merchants certainly love gift cards. They get the cash up front without delivering any product other than a plastic card. Often a gift card creates higher revenues than the value of the card; for example, in the case of restaurants, a $20 card may lead to a $40 meal. And best for the merchant, some of the funds never get spent. A $20 gift card to a bookstore may get used for a $15 book and the balance may never be exchanged for goods or services. The difference—that unspent balance—is 100 percent margin for the retailer. Consumers love them. Merchants love them. Sounds like a business.

---

102 Wreden, Nick. (2004). *Wireless: The Next Generation.*
103 Cannon, Ellen. (2007, November 12). *2007 Gift Card Study.*

CONCLUSION

I carry a credit card that earns frequent flyer miles on my preferred airline. I charge almost everything I can to this card to earn miles. I pay the whole bill off every month so I never have to pay any interest. There are no lemonade stands in my neighborhood, so I have little need of cash and don't spend much. But I usually carry about $100 in cash anyway. The only recurring need I have for cash is my weekly poker game with a $20 buy-in. Am I cashless?

Unlike a paperless society, my definition of a cashless society does mean *no* cash. Of the three requirements that must be met to move to a completely cashless society—security, privacy, and ease of use—credit and debit cards offer two out of three, failing only the privacy test. But I don't use my credit card for my weekly poker game out of a fear of loss of privacy; I don't use it because the folks I play with have no means to credit or debit it at the end of the evening. Even discounting the lack of privacy, credit cards aren't the same as cash because only merchants can credit your card. Credit cards are meant to spend money, not receive it.

My vision of the Swiss Army pocket knife for our lives—a mobile device that will look very much like today's mobile phone—will solve the how-do-I-get-my-money issue. People will deposit money into the account associated with my mobile device, and I can transfer money to others from that account. Think about the cashless chain today: your employer puts money into your checking account via direct deposit and you use that account to pay bills, often electronically, which is the way I pay my credit card every month. Using a mobile phone to promote a cashless economy will allow this to happen without a bank's direct involvement. And it will facilitate smaller transactions, like buying a soft drink from a vending machine, and traditionally cash transactions, such as tipping the pizza delivery boy. Interestingly, Japan and Hong Kong are now reporting declining levels of small currency in circulation.[104] There is no reason to think it won't happen here too. Most of us put pennies in a jar rather than in our pocket.

Privacy remains the biggest stumbling block to a cashless society, and the key to its success or failure. Cash is great for buying your spouse a present

---

104 *Mobile Payments in Asia Pacific.* (2007). p. 19.

that you don't want to appear on a bill and ruin the surprise. And cash is great when you want to buy a soft-porn movie and might face questions later from senators (who have likely done the same) voting to confirm your nomination to serve on the Supreme Court. Privacy was a critical success factor in the sale of videocassette players—people could watch what they wanted to in the privacy of their own homes. The same is true for early Internet users—pornography was one of the first moneymakers, and the privacy of the medium drove that revenue. Online gambling was also an early moneymaker on the Internet, though perhaps convenience and availability were bigger drivers than privacy. But privacy concerns always play a role in commercial activities where the public doesn't have a common view of right and wrong. The long-term impact of privacy on the cashless economy seems clear: no privacy will mean little progress.

In October 2007, the Reverend Glenn Guest published *Steps Toward the Mark of the Beast*, which describes an "apocalyptic scenario following the Bible's Book of Revelation that begins with the gradual disappearance of cash and checks and the increased reliance on debit and credit cards."[105] If the Reverend is right, the end of days is near, because cash and checks are disappearing at much more than just a gradual rate. "Technology has progressed to the point where [a cashless society] can be implemented . . . [but] I could well be wrong on certain points," said Guest.[106]

Assuming, ever hopeful, that using a mobile device to buy a candy bar in Seoul is not actually fulfilling an apocalyptic prophecy, the success of the wireless revolution will lead to a successful cashless revolution. An e-wallet needs a home and the mobile phone is perfect, because it can communicate with the bank to put more money in and allow merchants to take out the appropriate amount after the user authorizes it with his biometric or coded approval. When the cashless revolution is complete, it won't seem like much of a revelation.

---

105 Starrs, Chris. (2008, February 6). Pastor Says Cashless Society
    Would Fulfill Bible Prophecy. *Athens Banner-Herald.*
106 Ibid.

WHAT IN THE WORLD IS WORLD IS GOING ON?

CHAPTER 8
THE REVOLUTIONS AROUND THE WORLD

The challenge for developing economies is not to get the unbanked to the bank, but to get the bank to the unbanked.[107]
—BRIAN RICHARDSON, CEO OF WIZZIT

There are only two kinds of people on Earth. When I studied political science in college in the early 1980s, we always talked about *us* and *them—us* being the developed world and *them* being the third world. Political correctness ended this nomenclature and political scientists started referring to third-world countries as *lesser developed countries* and later as *developing countries*. *Developing* sounds more hopeful, but actually did nothing to help these countries develop. What was really happening was that countries were clustering into three categories, those with a per capita gross domestic product of about $350 ($1 per day per person), $3,500 ($10 per day per person), and $35,000 plus ($100 per day per person or more).[108]

---

107 Bhengu, Xolile. (2007). *13 Million Can Cash In on Cell Phone Banking.*
108 The exact terms, *us* and *them*, and the $1/$10/$100 groupings are from Dr. Hans Rosling, Professor of International Health, Karolinska Institutet. His presentation on this topic was available as of May 1, 2008, from http://www.ted.com/index.php/talks/view/id/92.

Globalization and the collapse of communism—and the increased wealth it has brought to the former second world, China and India—is making the economic distinction more difficult to categorize. Now it can be argued again that there are only two kinds of people on Earth—those who have access to technology (e.g., the Internet, mobile phones) and those who don't. The richest countries—and their ability to develop and deploy technology—are getting richer still. The poorest countries, still struggling with subsistence agriculture and the crush of poverty, are getting further and further behind. The so-called *digital divide* is in fact a chasm that won't be crossed with $100 laptops, Wiki-based knowledge sharing, or other open-source software. Nonetheless, what is different about the wireless and Internet revolutions in developing countries as opposed to other changes (e.g., railroads and aviation)—is how much faster economies can integrate this technology into their culture.

Some 2006 statistics from the International Telecommunications Union (ITU) explain how wide the digital divide is, but also that some progress is being made to close it:[109]

- Just 5 percent of Africans used the Internet, compared with 50 percent in the G8 countries (Canada, France, Germany, Italy, Japan, Russia, the United Kingdom, and the United States).

- There are only slightly fewer Internet users in the G8 countries than in the rest of the world combined.

- There are five times as many Internet users in the United States, and two times as many Internet users in Japan, as there are on the entire African continent.

- The good news? Africa was the region that had the highest mobile-phone growth rate in 2006. Continent-wide, Africa had about two hundred million mobile-phone subscribers at the end of 2006.

I teach my students about the "four Cs" of e-commerce, one of the most powerful drivers toward the future of less. E-commerce needs computers, connectivity to the World Wide Web, credit, and confidence—confidence that the vendor will ship your product and the local postal system or package delivery service will bring it to your door. Mobile phones can replace

---

109 International Telecommunications Union (ITU) statistics. (2006).

computers and can provide at least some measure of data connectivity almost everywhere in the world. They can even serve as an alternative to credit (peer-to-peer payment systems), though the back-office hardware and software must be in place. But mobile phones can't change confidence and it is this cultural factor among others that is crushing economic development in so much of the developing world.

DISRUPTIVE TECHNOLOGY The rapid growth in e-commerce shows that people will embrace a disruptive technology when they understand the value. Instead of driving to the mall and walking for hours around all the stores, a few clicks and I have ordered what I want. Without the four Cs of e-commerce—computers, connectivity, credit, and confidence—this could not exist. Three of these—computers, connectivity, and credit—instantly exist in countries with reliable mobile-phone networks. The mobile phone can serve as a personal computer, can connect to a network, and can either be tied to a credit card or carrying a prepaid balance that can be spent like a stored-value gift card. Half of the fourth C—confidence—is thus assured: the merchant knows he'll be getting paid. The other half of confidence is the customer's need to know that the merchant will ship the product and the post office or delivery service will bring it to his house. That is something technology hasn't completely solved yet, but perhaps a rating system like eBay's where merchants earn confidence points for successful transactions is a place to start.

Nonetheless, many developing countries are already seeing some of the wireless, paperless, and cashless revolutions, with relatively robust wireless networks available in even the poorest countries. Mobile phones are among the first items purchased when disposable income exists and MNOs in developing countries are proving quite adept and signing up new customers using prepaid billing services. Commerce is already migrating to these phones in the absence of credit and bank accounts as peer-to-payments are meeting the needs of many of these kinds of consumers. Developing countries will benefit from the software technology produced and the hardware economies of scale achieved elsewhere—creating millions of new local customers, but no new local millionaires.

> This can be hard for people in the rich world to understand, because the ways in which mobile phones are used in the poor world are so different. In particular, phones are widely shared. One person in a village buys a mobile phone, perhaps using a micro-credit loan. Others then rent it out by the minute; the small profit margin enables its owner to pay back the loan and make a living. So although the number of phones per 100 people is low by rich-world standards, they still make a big difference.[110]

The three revolutions in developed regions/countries (e.g., Europe, Japan, Singapore, and South Korea) and those rapidly developing countries/regions (e.g., Brazil, China, India, Russia, Vietnam, and most of Eastern Europe) will have quite different results. In developed countries, wireless came about largely out of convenience. In developing countries, wireless was the only means of communication because the state-owned wired service provider could take months or years to install a phone in your home. There is plenty of evidence to suggest, and even more common sense to conclude, that the mobile phone is the technology with the greatest impact on development. A new paper finds that mobile phones raise long-term growth rates, that their impact is twice as big in developing nations as in developed ones, and that an extra ten phones per one hundred people in a typical developing country increases GDP growth by 0.6 percentage points.[111] Going paperless in developed countries means a way to reduce costs and increase the exchange of ideas. Unlike in the developed world, paperless in developing countries *can* mean no paper—no paper for books or schools. There the paperless revolution becomes a way to gain access to content, not just a means to share content more quickly. And while the cashless revolution in developed countries is about security, privacy, and convenience, in developing countries it may give the poor the only access they've ever had to basic banking services,

110 *Calling Across the Divide.* (2005).
111 *The Real Digital Divide.* (2005).

microcredit, and nonbarter exchange. All three revolutions are at work in developing countries—bringing poor people into the mainstream economy, promoting convenience (and thus increasing the quality of life), and allowing for personal savings, even without a bank account.

The digital divide that really matters, then, is not simply between those with a computer and those without, but between those who have access to a mobile network and those who don't. Economies of scale have forced down the price of mobile phones, resulting in a narrowing of the digital divide in many places. The United Nations has set a goal of 50 percent access by 2015, but a new report from the World Bank notes that 77 percent of the world's population already lives within range of a mobile network.[112] This is good news. Neither the government nor the UN will need a big marketing campaign to get the people in developing countries to use mobile phones; even the world's poorest people are already rushing to embrace mobile.

A PREVIEW OF THE FUTURE OF LESS: SOUTH KOREA TODAY

The future is now visible in only one country, South Korea . . . To see the future, you need to understand South Korea.[113] —*DIGITAL KOREA*

Perhaps the most fascinating part about the wireless, paperless, and cashless revolutions in South Korea is that it happened so quickly and had such a dramatic effect on the Korean economic miracle. Korea has no oil, no diamonds or gold or other precious metals, and no particularly productive farmland. Per capita GDP in South Korea was less than $100 at the end of the Korean War in 1953. Yet among the almost one hundred countries that became independent since then, none of them can come close to Korea's list of achievements. Korea today is on track to hit 20,000 in per capita gross domestic product (GDP) by 2010. It is the world leader in broadband access, and has a wireless telecommunications infrastructure that is one of the best in the world (and is three to five years ahead of that of the United States). With about fifty million people, Korea has a global ranking of twenty-four in terms of population, but

---

112 Ibid.
113 Ahonen, Tomi, and O'Reilly, Jim. (2007). pp. 3–4.

is ranked eleven in the number of telephone lines, seventeen in the number of mobile-phone users, seven for the number of Internet users, and number one in terms of broadband penetration (see figure 8.1).[114] It's not called the "Korean economic miracle" for nothing.

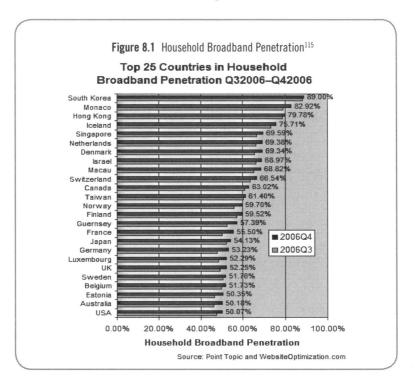

Figure 8.1 Household Broadband Penetration[115]

Top 25 Countries in Household Broadband Penetration Q32006–Q42006

Country	Penetration
South Korea	89.00%
Monaco	82.92%
Hong Kong	79.78%
Iceland	75.71%
Singapore	69.59%
Netherlands	69.38%
Denmark	69.34%
Israel	68.97%
Macau	68.82%
Switzerland	66.54%
Canada	63.02%
Taiwan	61.40%
Norway	59.70%
Finland	59.52%
Guernsey	57.39%
France	55.50%
Japan	54.13%
Germany	53.23%
Luxembourg	52.29%
UK	52.25%
Sweden	51.76%
Belgium	51.73%
Estonia	50.35%
Australia	50.18%
USA	50.07%

Legend: ■ 2006Q4 ▨ 2006Q3

Household Broadband Penetration

Source: Point Topic and WebsiteOptimization.com

Are Korea's state-of-the-art IT infrastructure and wireless networks the key to its economic success? Not solely, of course. There are important cultural factors including the Confucian focus on education, a strong work ethic, respect for authority, and long commutes that created a demand for state-of-the-art wireless networks. And we've seen similar success in other countries that have no natural resources other than the brainpower of their citizenry; for instance, Israel, Singapore, and Taiwan. But Korea's passionate embrace of technology and its ability to achieve world-leader status in so many technology-based vertical markets—mobile phones, flat panel displays, computer memory—is tremendously impressive.

---

114 All statistics and rankings are from the CIA World Factbook (https://www.cia. gov/library/publications/the-world-factbook/) as of August 28, 2007.
115 Point Topic and WebsiteOptimization.com. (2007, January).

HOW DID KOREA DO IT?

Where there is no tiger, the rabbit becomes king. —**KOREAN PROVERB**

Since so many nations of the world are facing an analogous, if not identical, steep climb to achieve economic viability and prosperity, South Korea's economic success may serve as a useful blueprint. First-world nations being leapfrogged on new technology should also take note: the new, global economy could leave them at a disadvantage to the upwardly mobile, rapidly developing countries.

South Korea's economic miracle started in 1963 with the rise to power of President Park Chung-Hee. Though obviously this predates Korea's rise as an IT powerhouse, President Park set forth economic policies—collectively known as state capitalism—that directly contributed to that rise. South Korea's state capitalism was based on the notion that free markets work, but that their biggest consequence—the demise of failed companies—must be minimized by activist government policies. For example, instead of having many companies manufacture cars with only the winners surviving, President Park's policies encouraged fewer market entrants and all but guaranteed market share at a rate that a firm could stay in business. It meant competition, but no big losers.

The government created the Ministry of Communications in 1984 and its expanding role was reflected in its 1994 name change to the Ministry of Communication and Information. In 1995, a government policy in the same spirit as President Park's turned out to be one of the most shrewd and far-sighted investments in business history. It spent big on a nationwide high-capacity broadband network that any wired network operator could offer service on, and offered subsidies so that forty-five million Koreans could buy cheap PCs. Cost: a mere $1.5 billion.[116] And this being Korea, the people bought PCs made in Korea. If you believe as I do that the collective elements of information technology—personal computers, the Internet, the World Wide Web, handheld electronic devices, and mobile wireless access—are among the most important inventions since the light bulb, it is easy for one to conclude that South Korea made building the number-one information technology country its Manhattan Project.

---

116 *The Future Is in South Korea.*

Building the #1 Information Technology Country

I think this picture of a large stone wheel says a lot about the role of the South Korean government in developing a domestic telecommunications industry. First, it sits in front of the Ministry of Information and Communications (MIC) in Seoul. The United States does not have such a cabinet-level office. The role of the Federal Communications Commission is largely regulatory rather than strategic. And the National Telecommunications and Information Administration, a bureau in the Department of Commerce, describes its fundamental role as the "President's principal adviser on telecommunications and information policy issues."[117] And although I have no idea if this was part of the symbolism, the fact that it is a wheel could demonstrate the MIC's understanding of what a telecommunications infrastructure is to a modern society—just

---

117 Available as of May 12, 2008, from the NTIA website,
   http://www.ntia.doc.gov/ntiahome/aboutntia/aboutntia.htm.

as the invention of the stone wheel more than years ago helped revolutionize virtually every aspect of human industry. Finally, the verb *building* speaks volumes too. The MIC is not just regulating or promoting policies; it played and continues to play a critical role in the creation in South Korea's technological development.

State capitalism is something that free-market Americans are not entirely comfortable with. The history of the United States is littered with failed firms—you are not driving a Hudson or a Nash or an AMC car. And our cultural distrust of the government's ability to do things better than the private sector rages on today with political battles over national health insurance. The United States does have its own flirtation with state capitalism, though, even if it is rarely consummated. Remember that the roots of the Internet are from U.S. government-backed research at DARPA in 1969. The United States is the only country on Earth where the government never owned the publicly switched telephone network. And following the 1982 break-up of AT&T, U.S. telecom firms have had to compete, with the losers being forced to exit the market. This happens less in South Korea, though the 1997–98 economic collapse did force many smaller firms into bankruptcy.

President Park and the South Korean government would likely argue that as a small, poor country in the early 1960s, Korea couldn't afford the luxury of letting the marketplace choose winners and losers. So the Korean people were initially left with a handful of firms dominating the telecom market, most notably the then 100 percent government-owned Korean Telecom. The 1990s saw the beginnings of true competition in wireless, with five national MNOs competing. The leading South Korean MNO, SK Telecom, later bought Shinsegae and KT Freetel later bought Hansol PCS, leaving those two and LG Telecom.

The South Korean government did have two policies that increased competition. Unlike in the United States, where each home or business must choose its traditional long-distance service provider for all such calls, in South Korea customers can make that decision on a call-by-call basis by choosing a different prefix, e.g., 001+number, 002+, 008+. And wirelessly, all MNOs also had their own prefixes to the numbers customers had, 011-xxxx-xxxx, 015-, 017-, 018-, and 019. This meant that if a customer wanted to change service providers, that customer had to get a new phone number. Now all new numbers are assigned a 010 prefix, so customers can choose any MNO and keep that number. Further, given the absence of GSM technology, a single-mode (CDMA) dual-band phone would not need to be replaced if a consumer changed service providers.

One of the most interesting comparisons between the development of wireless networks in the United States and South Korea is how frequency is allocated by their respective federal governments. The United States has auctions where potential service providers must spend billions to secure frequency. This expenditure precedes renting lower sites, putting up the first tower, and marketing the new service. U.S. service providers are therefore slower to roll out networks and new services since so much of their capital (or, more often, their access to credit instruments) is tied up in obtaining the rights to use the frequency.

In contrast, South Korea makes the frequency available to service providers at a greatly reduced rate using the argument that airwaves are a public good, like parks and national defense. In exchange for access to the frequency, service providers are required to roll out nationwide networks in a certain timeframe. This results in better coverage and more services available sooner. Of all the policies, it is this that has made Korea a world leader in the wireless revolution.

If the notion of this kind of government action seems opposite to traditional U.S. public-private barriers, remember the early days of television. The U.S. government essentially gave away tremendously valuable frequency in the 700 MHz band to three main broadcasters: ABC, CBS, and NBC. In exchange, these broadcasters had to follow just a few simple rules about what language and content was appropriate at various times of the day (e.g., *the family hour* and *the seven dirty words*) and days of the week (e.g., religious programming on Sundays).

In short, it was not so long ago that South Korea was a developing country; now it is one of the largest economies in the world. There is hope for some other countries to make similar progress. Even if some regions fail to emerge, the wireless, paperless, and cashless revolutions will still positively impact the poorest of the poor in ways we cannot yet imagine.

LESS IN CHINA

The Chinese people seem to be way ahead of Americans
in living a digital life.[118] —**BARRY DILLER**

Given a population of over 1.3 billion people, all the numbers for China are
huge. China has more than four hundred eighty million wireless subscribers
(about equal to all of Europe's subscribers), one hundred fifty million people
online, and will soon overtake America as the country with the world's largest number of Internet users. British pop star Katie Melua recorded a song
with the lyrics "There are nine million bicycles in Beijing." And Starbucks,
with about two hundred fifty stores in China, is supposedly opening a new
store every day.

Compared as it often is to India, the only other nation with a billion
people, China is putting up better numbers. Lehman Brothers calculates that
in 2007, China had 72 computers per 1,000 people; India had 24. China had
more than 50 million broadband connections; India had only 2.6 million,
using the broadest possible definition of *broadband*.[119]

> [In 2007] China's online population was estimated at 137 mil-
> lion people, second only to the United States at 165–210 million.
> According to the *Young Digital Mavens* survey, 80% of young Chi-
> nese people believe that "digital technology is an essential part of how
> I live," compared with 68% of Americans. Twice as many Chinese as
> Americans (25% to 12%) said they would not feel okay going with-
> out Internet access for more than a day.[120]

The mobile wireless market has been one of the fastest-growing sectors
in China, where the mobile infrastructure is comparatively much better than
the wired infrastructure. It is not unreasonable to suggest that most Chinese
alive today have never made a phone call over a traditional wired network and
never will. These unwired Chinese are the reason that the Chinese mobile
market is booming. Those 480 million wireless customers are served by just

---

118 *America's Emobyte Deficit.* (2007).
119 *Does Not Compute.* (2007).
120 *America's Emobyte Deficit.* (2007).

two MNOs, China Mobile and China Unicom. Two MNOs means that even with almost half a billion customers, it is relatively easy to roll out new services to lots and lots of captive customers. An estimated 4 million Chinese become new wireless subscribers every month.[121] And they are looking for new things to do with their new phones.

The availability and success of e-commerce in a developing country is a good way to evaluate progress regarding the wireless, paperless, and cashless revolutions. E-commerce is supported by the wireless revolution and supportive of the paperless and cashless revolutions. And as in all countries, the challenge of e-commerce in China rests firmly on those four Cs—computers, connectivity, credit, and confidence. In China, the prospects for e-commerce are improving quickly. Handheld mobile devices are becoming the computers and China's MNOs provide the connectivity. The number of credit cards in China jumped to more than forty million in 2006 from fewer than three million in 2003, according to the consulting firm McKinsey.[122] And even without a credit card, Chinese MNOs permit a mobile user to prepay funds that can be used in lieu of credit.

Confidence—merchants trusting customers and customers trusting merchants—remains the biggest obstacle to more e-commerce in China. Moreover, Chinese consumers are personally liable for fraud and misuse of credit cards. Combined with lack of trust between buyer and seller, this means that Chinese consumers are not yet realizing all the benefits of e-commerce and mobile e-commerce. Cash-on-delivery remains a dominant form of payment even in the major cities.

In the long-term, mobile payments in China will be successful because the Chinese government views the development of a low-cost, noncash payments network in rural areas as critical to increasing rural spending and closing the wealth gap with urban areas, according to a 2007 study by the global consulting firm KPMG.[123] China is therefore using its wireless network to improve the rural quality of life in ways other than just increased communications.

One homegrown solution to the distrust and risk inherent in Chinese e-commerce is Yeepay (www.yeepay.com/aboutus/), which has targeted the

---

121 *Convergence Takes Hold.* (2007). p. 26.
122 *China to Relax Credit Card Market.* (2007, December 19).
123 Ibid., p. 14.

confidence gap in the face of growing demand. According to their website, Yeepay "provides an integrated payment platform that enables any consumer or business in China to send and receive payments on the Internet, telephone, or mobile phone securely, conveniently and cost-effectively." In 2006, for example, the Chinese government decreed that all domestic airline tickets become paperless. Customers can now purchase tickets over their phone and then check in simply by displaying their ticket details on their mobile phone. Yeepay has gone after the corporate domestic travel sector. This wireless e-ticket market is estimated at $15 billion for 2008.[124] For Chinese citizens without credit, there is a strong network of government-run rural post offices that can act as agents for cash deposits to prepaid phone accounts.

One dangerous trend in China—dangerous in that it threatens to dramatically slow down wireless growth and the applications that need wireless to grow—is the government's interest in making China-only proprietary products. The idea is that if the specifications for a product mean it will work in only one country, companies in other countries won't develop a product for it and local manufacturers will have 100 percent market share. For a small country or a small industry, that is okay. But China is a huge market and the people will be best served by lots of competition, which will lower costs and improve the quality of products available to Chinese consumers.

> Not content with DVD, the government introduced its own format for video discs, EVD. Unhappy with MPEG-4, it created an alternative standard for compressing audio and video files called AVS. Dissatisfied with the Wi-Fi encryption method that everyone else used, it tried to push an alternative called WAPI. China's latest gift to the world is Time Division-Synchronous Code Division Multiple Access (TD-SCDMA). Developed by China's Datang and Germany's Siemens, this allows mobile devices to send and receive bigger gobbets of information, including video and Web pages. Two alternatives are already widely used worldwide. Why abuse the alphabet further?[125]

---

124 Ibid., p. 23.
125 *Consumer Champion.* (2007).

The consequence of this policy is that China is way behind almost everyone else in Asia in terms of deploying 3G networks. As *The Economist* politely jabs, "China makes many of the world's 3G phones, but almost none of the world's 3G phone calls."[126] I predict that China will eventually join the global wireless standards–based networks, or at least allow them to compete with China's nonstandard networks.

LESS IN INDIA

As one of the fastest-growing wireless markets in the world, the wireless subscriber base in India has witnessed exponential growth since the year 2000. The total subscriber number is expected to exceed 260 million by 2010 and generate total revenue of $17.5 billion.[127] —INSTAT

India, with over 1.1 billion people, has the fourth- or fifth-largest GDP in the world, but ranks about 150th on a per-capita basis, less than a third of the global average.[128] The country may be prospering, but its people are desperately poor. There are about two hundred million households, or eight hundred million people, who have no access to banks or formal financial services. Many of the unbanked are migrant workers who want to send money back to their families.[129]

Certainly, not everyone in India is poor. The country is a long-standing member of Asia's million-a-month club for its rate of new mobile subscribers, along with China, Pakistan, and Indonesia. The demand for ultra-low-cost mobile phones, those selling for less than $75 each, gets a lot of attention. But InStat reports that 33 percent of over 1,000 primarily urban Indian respondents to a 2007 survey said that they would not even consider a basic mobile handset with only voice and SMS capability.[130]

---

126 Ibid.
127 Retrieved May 12, 2008, from http://www.instat.com/catalog/apcatalogue.asp?id=236.
128 Retrieved May 12, 2008, from https://www.cia.gov/library/publications/
    the-world-factbook/rankorder/2004rank.html.
129 *Mobile Payments in Asia Pacific.* (2007). p. 14.
130 Retrieved May 12, 2008, from http://www.instat.com/catalog/apcatalogue.asp?id=236.

So India has both extremes—rich and poor—and a huge middle class. What all three groups seem to have in common, at least statistically, is a tendency to talk a lot. The average owner of a mobile handset (in India) spends 471 minutes (almost eight hours) on the phone each month and sends 39 text messages, but that doesn't mean a lot of revenue for Indian MNOs, which earn under $7.50 a month on average per subscriber.[131]

During the dot com boom in 2000, a global network of undersea fiber optic cables were deployed that created the infrastructure that led to India's ability to, for example, serve as a call center for many U.S. firms. It is this technology that created much of the new national wealth in India, though it is highly concentrated in the southern part of the country. Now technology offers an opportunity to help raise the quality of life for the almost billion people in India not living in the southern cone who are not employed by India's wealthy outsourcing firms.

The ongoing wireless revolution in India means that mobile phones are making life better for people in remote, underserved, largely rural areas of India. Even some of India's poor no longer have to walk to neighborhood public call offices to make a call. With more than two hundred fifty million mobile users and six million new ones added each month, Indian MNOs can now support more sophisticated mobile technologies. An extra ten mobile phones per one hundred people in a typical developing country leads to an additional 0.59 percent of growth in per capita GDP, according to a London Business School study.[132]

The key to success in such a market will be customization—products designed primarily for the Indian market, or at least focused on developing economies in general. Customization will mean inexpensive mobile phones, creative accessories, and innovative applications. Low-price mobile devices are already available and increased manufacturing or final assembly in India could lower that price further still. In January 2007, Motorola announced a phone recharger that works while the owner is riding his bike.[133] But it is the newly available services that will have the most impact on India's poor.

---

131 *Does Not Compute.* (2007).
132 Overdorf, Jason. (2007). *Cashless in the Hinterlands.*
133 McLaren, Warren. (2007). *Motorola's Bike Charger for Mobile Phones.*

Several small companies are already at work on mobile banking for small businesses in India, among them Ekgaon (www.ekgaon.com), which has developed a system for tracking transactions for groups of individuals who offer loans to poor people.[134] Mobile banking services can reduce the cost of transactions for loans and other services—the main obstacle to providing banking for the poor—by as much as three quarters, according to Ekgaon's chief operating officer. By March 2008, people in 8,000 villages in Andhra Pradesh will get their social security payments sent to them via their mobile phones, which they may eventually be able to use instead of cash.[135]

Certainly India's growing middle class offers tremendous market opportunities beyond just entry-level mobile devices and micro-payments. These wealthier customers want more—and can afford better—services. And the overall poor quality of India's wired infrastructure is a driving force behind the rush for wireless services. The number of Indian consumers connecting to the Internet via mobile phones more than doubled to thirty-eight million in 2006, according to a report by the Telecom Regulatory Authority of India.[136] Accessing the Internet on mobile phones has been so successful that it has exposed how weak India's traditional Internet networks are. This is fueling a race for customers and sales between the country's MNOs and its wired service providers. Some experts attribute the surge in wireless Web use to a combination of falling handset prices, network upgrades, and an economic expansion that's leaving many young people flush with disposable income.[137]

Unlike Japan and South Korea, however, the wireless networks in India are still relatively expensive to use and thus not yet likely to hasten the migration from wired to wireless, even with the generally poor wired infrastructure. Those struggles notwithstanding, demand for wireless Internet access is likely to keep skyrocketing. The Indian Cellular Association expects two hundred million people to sign up for mobile Web access by 2010.[138] Do the math: two hundred million people using just ten minutes a day means sixty billion mobile minutes online each month. As we've seen in China, when the law of really big numbers kicks in, you see some really big numbers.

---

134 Overdorf, Jason. (2007).
135 Ibid.
136 Lakshman, Nandini. (2007). *A Wireless Revolution in India.*
137 Ibid.
138 Ibid.

LESS IN THE PHILIPPINES

There is plenty of evidence that once people gain access to a phone,
they find many ways to exploit it to their benefit.[139] —**KPMG STUDY**

Although the mobile market in the Philippines is much smaller than that
in China or India, the wireless revolution and the migration to mobile (and
therefore cashless) banking is already happening there. One reason is that
about 8–10 million Filipinos work overseas and send money home, equivalent to 10 percent of the country's GDP.[140] Such remittances are simply one-
way, peer-to-peer transactions.

> While Japanese and South Korean consumers have been using cell
> phones as virtual wallets for several years, those systems use a com-
> puter chip implanted in [the] handset that allows people to buy things
> by waving the phone in front of a sensor. The Philippine system relies
> on simple text messages, which cost just $.02 to send. Mobile phone
> users in the Philippines have embraced text messaging. The electronic
> connections have fostered a culture of quick greetings and forwarded
> jokes. Text messages also played a key role in mobilizing crowds that
> fueled the 2001 people power revolt that ousted President Joseph
> Estrada. The Philippines' two biggest MNOs have harnessed this
> penchant for text messaging to enable consumers to enter the world
> of e-commerce.[141]

The standard international remittance method—bank-to-bank wire
transfer or using a firm like Western Union—costs about $2.50 and takes
two days to clear. But a mobile phone doubling as an e-wallet that can send
and receive money via text messaging can cut that cost by 90 percent, offer
near instantaneous transfers, and work without a bank account to retrieve
the money. More than 5.5 million Filipinos now use an e-wallet, making the
Philippines a leader among developing nations in providing financial transactions over mobile networks.[142]

---

139 *Mobile Payments in Asia Pacific.* (2007). p. 26.
140 Ibid., p. 22.
141 Teves, Oliver. (2007). *Cell Phones Double as Electronic Wallets.*
142 Ibid.

Customers can deposit and withdraw cash through an MNO's prepaid sales agent and send money to other people via text messages, which can then be exchanged for cash by visiting any other agent. Workers can then be paid by phone; taxi-drivers and delivery drivers can accept payments without carrying cash around; money can be easily sent to friends and family. A popular use is to deposit money before making a long journey and then withdraw it at the other end, which is safer than carrying lots of cash.[143]

The next stage will be to combine m-banking services and microfinance loans, extending access to the unbanked. Some MNOs already issue customers with debit cards linked to their m-banking accounts, just the opposite of the trend in more developed countries. All this has the potential to give the poor more access to financial services and bring them into the formal economy.

GLOBAL CONSUMERS Your business is almost certainly global by some definition—you either have international suppliers or international customers. Perhaps you have outsourced some part of your manufacturing to China or your customer service to India to save on labor costs. If you are fortunate, you have global customers—fortunate because more than 95 percent of the world doesn't live in the United States.

So it matters what *they* are doing over there. They are your workforce and your future customers. This might mean they are using some other company's product today. They are already part of the wireless, paperless, and cashless revolutions. Does your current way of doing business offer the promise of winning them over? Just as you may have already changed some hiring practices and workplace rules to accommodate the new generation of American workers, so too should you consider what changes you need to make to accommodate your future global customers. It is a matter of understanding the culture in both cases.

---

143 *A Bank in Every Pocket.* (2007).

In December 2000, the first mobile-banking service, SMART Money, was launched in the Philippines. In 2007, SMART expanded its m-payment service to include international remittances. SMART Money was the world's first reloadable e-wallet, linked together by a cellular network. Once cash has been transferred to the SMART Money account, it can be used in thousands of shops and restaurants. The cash value can also be used to load airtime, pay utility bills, or transfer money from one SMART Money card to another.[144]

A second service, G-Cash, was launched in October 2004 and initially focused on three types of transactions: international and domestic remittances, phone-to-phone transfers, and payments for retail purposes. Unlike SMART Money's approach, G-Cash's parent maintains records of all transactions and arranges settlement between the retailers and G-Cash customers. G-Cash provides services through about 4,900 retail outlets nationwide and more than five hundred G-Cash partners.[145]

As an example of how it works in the Philippines, imagine a son is off to a bank to collect his weekly allowance, sent by his mother—who's working in Hong Kong—to the electronic wallet in his cell phone. He walks into a branch of his local Philippine bank, fills out a form, and sends a text message on his phone to a bank line dedicated to the service. The transaction is approved in seconds and the teller gives him his money, minus a 1 percent fee. He doesn't need a bank account to retrieve the money.[146]

A look at what is happening in the Philippines is instructive for several reasons. First, this is a poor country with poor people. The per capita GDP is about $3,300, or about 1/14 of the United States'. Unlike Japan, South Korea, or Singapore, the Philippines cannot throw a lot of money at the problems the three revolutions are solving elsewhere. And its average education level compared to the Asian economic tigers isn't exactly laying the foundation for the Philippines to catch up anytime soon. But in the "if they can do it there, then we should be able to do it here" mantra, the Philippines can be a role model for the poorer countries in southeast Asia and Africa.

What is going on in the Philippines is also instructive because the success of mobile banking there is a classic example of local technology solving local problems. The local approach to mobile banking is ingenious and working

---

144 Amin, Shaker. (2007).
145 Ibid.
146 Teves, Oliver. (2007).

well. A huge source of hard currency in the Philippines is what Filipinos working outside the country send home. They had to develop something more secure than hiding hundred-dollar bills in letters and something less expensive than using Western Union.

LESS IN AFRICA

Do people have a clear view of what it means to live on $1 a day? About 99 percent of the benefits of having a PC come when you've provided reasonable health and literacy to the person who's going to sit down and use it.[147] —**BILL GATES**

It often seems that Africa is less an emerging market than a submerging one. Despite tremendous natural resources and forty-plus years of independence in some countries, no country in Africa can compete with the resource-starved success stories in the Pacific Rim. The Asian Tigers—Hong Kong, Japan, Taiwan, Singapore, and South Korea—chose the path of rapid migration from agriculture to high-tech industrialization and now enjoy the great wealth that decision fostered.

Only 20 percent of Africans have a bank account.[148] As of 2006, about 22 percent of Africans have a mobile phone (see figure 8.2).[149] But according to trends from the International Telecommunications Union, it appears that the rate of mobile-phone growth will accelerate much faster than traditional bank account growth (see figure 8.3).

As we saw in the Philippines, local problems often require local solutions. The challenge in much of Africa is giving people a way to exchange money that is safe and secure, but still allows them to be part of the global supply chain.

In Zambia's capital city, Lusaka, between stalls of dried fish, a red shipping container is loaded with Coca-Cola bottles. The local distributor for the market sells all its stock every few days. A full load costs about $2,000. In cash, this amount can be hard to get hold of, takes ages to count and—being

---

147 *Behind the Digital Divide.* (2005).
148 *On the Frontier of Finance.* (2007).
149 International Telecommunications Union (ITU) statistics. (2006).

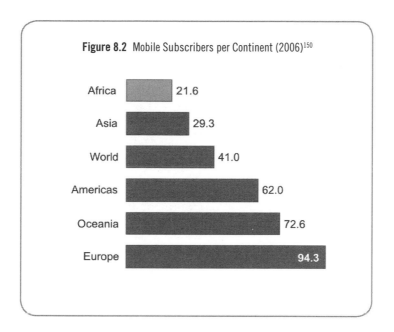

Figure 8.2 Mobile Subscribers per Continent (2006)[150]

Africa 21.6
Asia 29.3
World 41.0
Americas 62.0
Oceania 72.6
Europe 94.3

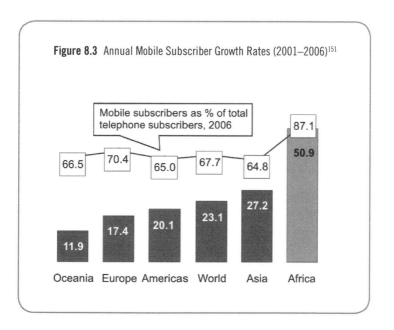

Figure 8.3 Annual Mobile Subscriber Growth Rates (2001–2006)[151]

Mobile subscribers as % of total telephone subscribers, 2006

Oceania 66.5 / 11.9
Europe 70.4 / 17.4
Americas 65.0 / 20.1
World 67.7 / 23.1
Asia 64.8 / 27.2
Africa 87.1 / 50.9

150 ITU. (2006). Retrieved May 12, 2008, from http://www.itu.int/ITU-D/ict/statistics/ict/graphs/ af3.jpg. Reproduced with the kind permission of the ITU.
151 ITU. (2006). Retrieved May 12, 2008, from http://www.itu.int/ITU-D/ict/statistics/ict/graphs/ af2.jpg. Reproduced with the kind permission of the ITU.

ten times the average annual wage—is tempting to thieves. So Coca-Cola now tells its three hundred Zambian distributors to pay for deliveries not in cash, but by sending text messages from their mobile phones. The process takes about thirty seconds, and the driver issues a receipt. Faraway computers record the movement of money and stock. Coca-Cola is not alone. Around the corner from the market, a small dry-cleaning firm lets customers pay for laundry using their phones. So do Zambian petrol stations and dozens of bigger shops and restaurants.[152]

South Africa, which in 2007 had about a 60 percent mobile penetration rate, is a mobile banking leader on the continent, thanks in part to innovative service providers like Wizzit (www.wizzit.co.za), a division of the South African Bank of Athens. But South Africa's banking system and mobile-phone penetration rates are both statistical outliers compared to the continent as a whole, which is generally way behind the rest of the world. Most banks in Africa have branches only in urban areas. As a result, regular bank services are often simply unavailable. Ethiopia has just one bank branch for every 100,000 people—compare that with Spain, which has ninety-six branches for every 100,000 people. Moreover, requirements to maintain relatively high account balances make banking services too costly for most Africans.[153]

Some counterparts to Wizzit have emerged elsewhere in Africa. Kenya has M-Pesa (www.safaricom.co.ke/m-pesa), which, unlike Wizzit, is run by the MNO and not a bank, so no bank account is needed.[154] In Kenya, 17 percent of those who are unbanked own a mobile phone. Financial services are offered over mobile phones, which serve more like electronic savings accounts than banks. Mobile subscribers can open accounts, check their balances, pay their bills, and transfer money via their mobile phones. In Kenya, where three million people have bank accounts, as many as one million use a mobile-payment service.[155]

Similar services are now available in Botswana, the Democratic Republic of the Congo, and Zambia. "The greatest impact is in rural areas," says an officer for Wizzit, "where 80 percent of all farmers do not have bank accounts."[156]

---

152 *Calling Across the Divide* (2005).
153 Kimani, Mary. (2008). *Africa: A Bank in Every Pocket?*
154 Retrieved May 12, 2008, from http://www.safaricom.co.ke/index.php?id=228/.
155 *On the Frontier of Finance.*
156 Kimani, Mary. (2008).

Africa's biggest challenge is—and will likely remain—corrupt and inept governments. And as we saw in the rapid IT development of Japan, Singapore, and South Korea (actually, Asia in general—see figure 8.4), an enlightened government and a constructive regulatory environment promotes innovative services and eager service providers. Thus this model would be beneficial to both developing and third-world countries. To paraphrase Garrett Hardin, good laws make us more free, not less.[157]

The regulatory approach being taken in the Philippines provides a good model for other countries. Rather than trying to work out the best rules in advance, which could hamper innovation, the regulator is working closely with the banks and operators behind the country's two m-banking services. The experience will feed into new banking regulations. Rules that are too tight will hinder adoption; rules that are too lax could allow fraudsters to bring the whole idea of branchless banking into disrepute. But if regulators strike the right balance, m-banking may provide the next example of the mobile phone's transformational power.[158]

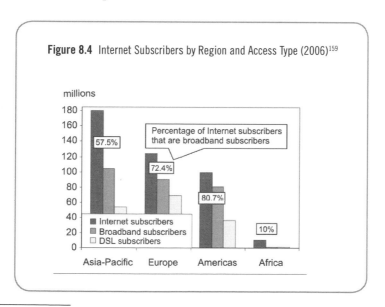

Figure 8.4 Internet Subscribers by Region and Access Type (2006)[159]

---

157 Hardin, Garrett. (1968). *The Tragedy of the Commons.*

158 *A Bank in Every Pocket.*

159 ITU. (2006). Retrieved May 12, 2008, from http://www.itu.int/ITU-D/ict/statistics/ict/graphs/am5.jpg. Reproduced with the kind permission of the ITU.

While it can be argued that mobile banking may not be the most necessary factor for an improved banking service in developing countries, it is certain that traditional banking service alone may not result in any significant improvement in providing the unbanked community with access to the financial sector. The speed and efficiency with which money can be transferred and monitored through such mobile platforms is likely to be far greater and higher as compared to a cash-based system.[160] And this includes countries with unstable government and little infrastructure, where these platforms offer the promise of serving the citizenry much better.

ONE LAPTOP PER CHILD OR LESS?

Encouraging the spread of mobile phones is the most sensible and effective response to the digital divide . . . The debate over the digital divide is founded on a myth—that plugging poor countries into the Internet will help them to become rich rapidly.[161] —*THE ECONOMIST*

The One Laptop Per Child (OLPC) project—formerly and less accurately known as the "hundred-dollar laptop"—is the brainchild of its chairman, Nicholas Negroponte, the co-founder and director of the famed MIT Media Lab. OLPC's mission is "to provide children around the world with new opportunities to explore, experiment, and express themselves."[162] In simple terms, OLPC's plan is to make a low-cost laptop to be purchased by governments and distributed to children. According to OLPC, that laptop, named the XO,

> . . . is a potent learning tool created expressly for children in developing countries, living in some of the most remote environments. The laptop was designed collaboratively by experts from both academia and industry . . . The result is a unique harmony of form and function; a flexible, ultra-low-cost, power-efficient, responsive, and durable machine with which nations of the emerging world can leapfrog

160 Amin, Shaker. (2007).
161 *The Real Digital Divide.*
162 Retrieved May 12, 2008, from http://www.laptop.org/en/vision/index.shtml.

decades of development—immediately transforming the content and quality of their children's learning.[163]

OLPC's five principles are:[164]

- Child Ownership (and free use)
- Low Ages (6–12 years old)
- Saturation (village, municipality, region, country)
- Connection (mesh networking)
- Free and Open Source (no inherent external dependency)

Due in large measure to Negroponte's charisma and passion for the project, OLPC has generated substantial media attention. Rising costs, delayed availability, and second thoughts from some of the would-be government buyers have taken some of the shine off the project, though the idealism remains. In December 2007, I participated in the "Give One Get One" program, where I bought one XO for about $400, which also funded the distribution of another laptop to a child in a developing country.

I am outside the demographic of an intended user, and my main reason for buying an XO laptop was educational, not charitable; I wanted to show it to my students. Even so I'm now less optimistic about the long-term success of OLPC. Given the tremendous growth in the number of mobile subscribers—even in the poorest countries in Africa—I still argue that the mobile phone is the Internet access device of choice for the developing and third world.

> Handing out $100 laptops to kids in rural African villages is great, but it won't bridge the digital divide. That's because it's so hard to find reasonably priced, high-speed Web access. In Japan, DSL or cable averages 6 cents per 100 Kbps. In Kenya, that same hook-up speed costs $86.11—nearly twice the average monthly income.[165]

The business model of having governments buy the laptops and distribute them at no charge to children in their countries is reasonable and practical.

---

163 Retrieved May 12, 2008, from http://www.laptop.org/en/laptop/hardware/.
164 Retrieved May 12, 2008, from http://wiki.laptop.org/go/Core_principles/lang-en.
165 *Access Denied*, p. 60.

And this strategy seemed largely intended to reach economies of scale more quickly so as to reach the approximate $100 price target. But handheld mobile phones are already well under $100 and, with some slight modifications to the screens, could increase readability outdoors. The biggest upside to mobile phones is that they connect to existing networks, which already have a well-defined, proven bandwidth growth plan. And there is plenty of evidence to suggest that mobile-phone technology will have the greatest impact on development. A new paper finds that mobile phones raise long-term growth rates, that their impact is twice as big in developing nations as in developed ones, and that an extra ten phones per one hundred people in a typical developing country increases GDP growth by 0.6 percentage points.[166]

Further to this point, OLPC relies largely on Web-based software, rather than big hard drives with all the needed software downloaded to it. Since a handheld mobile device can easily replicate this functionality, the added cost, bulk, and less portability of an XO laptop seem to put it at a huge disadvantage in terms of gaining broad market acceptance. Mobile phones are already cheaper than personal computers, and companies like Nokia and Motorola are focused on making new kinds of phones that are even less expensive. There are at least three times more mobile phones than there are personal computers and the growth rate for phones is twice that of personal computers. In developing countries, the target for the OLPC initiative, it is generally easier to access the Internet with a mobile phone through the local cellular network than it is with a personal computer via the wired Internet. What is more, the World Bank estimates that more than two-thirds of the world's population lives within range of a mobile-phone network.[167] Mobile phones are already the next big Internet phenomenon in developing countries because they offer greater access for more people, with all the benefits that access entails.

The OLPC model—selling the computers to governments who in turn give them away—is a creative model that could succeed, though the number of countries signing up to date is small and getting smaller as some back out from previous commitments. Much to the chagrin of the OLPC project, it didn't scare away all competitors. Intel, an early partner in OLPC (even though the first version of XO ran on Intel's arch rival AMD's processor),

---

166 *The Real Digital Divide.*
167 De Waele, Rudy. (2006).

developed a competing product called the Classmate PC. It is hard to say if Intel's entry into the sub-$500 laptop market precipitated OLPC's decline or was in response to their declining fortunes. It is possible, perhaps likely, that Intel simply didn't want to cede a potential market of hundreds of millions of units to an AMD chipset.

Intel markets its Classmate PC as follows:

> The World Ahead Program from Intel aims to enhance lives by accelerating access to uncompromised technology for everyone, anywhere in the world. Focused on people in the world's developing communities, it integrates and extends Intel's efforts to advance progress in four areas: accessibility, connectivity, education, and content. The Classmate PC is a revolutionary new device targeted at providing one computing solution per student in emerging markets, taking advantage of the education focus to deliver a product that provides great student education in a rugged industrial design intended for children.[168]

In January 2008, Intel dropped out of the OLPC program, citing disagreements with the organization. "OLPC had asked Intel to end our support for non-OLPC platforms, including the Classmate PC, and to focus on the OLPC platform exclusively," an Intel spokesman said. "Intel believed all along that there is a need for multiple alternatives to meet the needs of children in poor countries."[169]

Time will tell if OLPC or Intel comes to dominate the youth market in the developing world. But the broader issue of whether low-cost laptops will actually bridge the digital divide is certainly more critical than whether a laptop or a handset is a more appropriate piece of hardware. On this point, the future doesn't seem bright. The digital divide is a symptom of other divides of income, access to capital, and literacy. In poor countries, not surprisingly, fewer people own computers and have access to the Internet simply because they are too poor, illiterate, or focused more on basic subsistence. So even if it were possible to wave a magic wand and cause a computer to appear in every household, or create wireless networks that made existing mobile phones more universally

---

168 Retrieved May 12, 2008, from http://www.intel.com/intel/worldahead/classmatepc/.
169 Wong, May. (2008). *Intel Quits One Laptop Per Child Program.*

Internet-ready, it would not completely bridge the digital divide. A computer is not useful if you have no food or electricity and cannot read.

CONCLUSION

The Mayans lost to the Aztecs, the Romans lost to the Barbarians, and the French lost to the Vietnamese. In each case the losers had superior technology, but the victors had more troops.[170]

—DUNDEE WEALTH MANAGEMENT

While researching this chapter on the developing countries, I came across something from the big thinkers at Frost & Sullivan that at first I thought contradicted my idea that the wireless revolution will enable the other two. "Mobile banking . . . will be a major driver for the adoption of wireless technology in rural areas in developing countries."[171] Certainly that statement is correct. Applications drive the purchase of hardware. No one buys a computer and says, "I think I'll send an e-mail." Rather, they know they want to send and receive e-mail, so they go out and buy a computer or a smart phone or whatever access-enabled device they choose. But I think I'm right, too. Once they have the computer they bought to send and receive e-mail, then they start to look at what else they can do with their new hardware. Web surfing comes to mind, in my example of the computer. Mobile banking comes to mind in the case of a mobile phone probably purchased primarily, or at least initially, because the user wanted to talk to other people.

What does all this mean? Consumers expect and will demand ubiquitous connectivity, but don't want to pay a lot for it. They value their social networks, but will leave and join new ones if given a reason. They know what they want and MNOs and content providers don't. It's not hard to see that the path toward the wireless, paperless, and cashless future will have a few bumps in the road.

---

170 Retrieved May 12, 2008, from http://whitepapers.zdnet.co.uk/0,1000000651,260013129p,00.htm.
171 Amin, Shaker. (2007).

AFTERWORD

Change is not merely necessary to life—it is life. —**ALVIN TOFFLER**

I wrote this book and you bought it, so I hope we agree that the wireless, paperless, and cashless revolutions matter today and will matter more in the years ahead. Wireless has already changed our lives in ways we never could have imagined just twenty years ago. Most Americans with a mobile phone would not trade it for any other consumer electronics product. And as the mobile phone morphs into the Swiss Army knife of our lives, it will be so many different consumer electronics products all rolled into one that there would be little to trade for.

The path toward a dramatic decrease in the use of paper is still possible. I hope that what seems to be a renewed interest in the environment helps more individuals and businesses go paperless faster. Even if your company isn't focused on green issues, then just do it for the certain cost savings you'll realize immediately.

The cashless revolution seems more certain than the paperless one, given the roll of credit, debit, and gift (stored-value) cards in developed countries and in prepaid cards in developing countries. Next-generation cashless payments will mean getting rid of those physical cards in exchange for a device—I'm guessing a mobile phone with biometrics—to greatly enhance the security of being cashless.

I'd like to conclude where I began, with a quote attributed to science-fiction author William Gibson: "The future is here. It's just not widely distributed yet." Our wireless, paperless, and cashless futures are here. The latest and greatest products and services are here—meaning they exist today—even if they are not yet available in the United States. They are coming. They will change your life. And in my opinion, they will change your life for the better.

APPENDIX

I n this book I discussed some of the implications of the wired Web 2.0 revolution on the wireless, paperless, and cashless revolutions. The wired Web will continue to evolve and is still way ahead of its mobile cousin. If you would like to learn a little more about what some companies are doing today in wireless, paperless, and cashless Web 2.0 space, I have compiled a short list of sites for you to explore. All of the descriptions provided were taken directly from the respective sites.

WIRELESS WEB 2.0

- **Bluepulse** (http://www.bluepulse.com/) "is the free mobile social messenger. Take the best of a social network, the best of an instant messenger, revolutionize both and combine them to run on anyone's mobile phone. That's a mobile social messenger."
- **ComVu** (http://www.comvu.com/) "is the world's first live video broadcast service for 3G mobile phones and other portable platforms. No more uploading. Instantly post videos to your own PocketCaster page,

blog, favorite video sharing or social networking site. Now, with a push of the button you can broadcast events to anyone in the world."

- **Loopt** (https://www.loopt.com/) "turns your phone into your social compass. Connect with friends and get alerted when they are nearby. Share your location, status and photos with friends and AIM buddies. Explore places and events recommended by friends."

- **Orb** (http://www.orb.com/) "manages all your media so you can consume it while you're away."

- **Plazes** (http://www.plazes.com/) allows users to "create activities to let your friends know what you are doing, when and where."

- **ShoZu** (http://www.shozu.com/) "connects your mobile phone to your favorite online social destinations."

- **SoonR** (http://www.soonr.com) "keeps you securely connected to your computer files so you're free to go mobile."

- **Twitter** (http://www.twitter.com) "is a service for friends, family, and co-workers to communicate and stay connected through the exchange of quick, frequent answers to one simple question: What are you doing? Why? Because even basic updates are meaningful to family members, friends, or colleagues—especially when they're timely."

- **WidSets** (http://www.widsets.com) allows users to "get your favorite Web content straight to your mobile phone. Find news and information, stay in touch with e-mail and blogs, play games, share pictures and more. WidSets uses mini-applications called widgets to push updates from your favorite sites directly to your phone."

- **Winksite** (http://www.winksite.com/) "makes it easy to create mobile websites and communities that can be viewed worldwide on any phone."

PAPERLESS WEB 2.0

- **Cellblock** (http://www.cellblock.com/) "enables instant multimedia publishing across websites, desktops and mobile devices (i.e., your cell phone). This means one or many people can be adding and viewing pictures and videos into a Cellblock via the Web, desktop, or mobile device. The instant part means that as soon as the content arrives in a Cellblock, it can be visible to everyone everywhere without page refreshes, new downloads or other impediments."

- **JuiceCaster** (http://www.juicecaster.com/) "lets you meet new people, connect with friends and share your photos and videos directly from your mobile phone. Instantly post videos and pictures to your MySpace, Facebook and other sites."

- **Phozi** (http://www.phozi.com/) "is a fun way to take photo booth pictures by yourself or with friends! Send in your photo, add graphics and drawings layered on top, then embed them into your social sites. You can send photos to your friends or post them anywhere on the Web."

- **Pickle** (http://www.pickle.com/) "was built from the start with both video and photos in mind. As a result, Pickle offers a truly integrated and complete solution for sharing personal media. As you use the video mode on your point-and-shoot digital camera or mobile phone more and more, you'll find yourself wanting to manage and share everything in one place. Pickle translates videos into a common, easily sharable format that just about all Web browsers can support."

- **PixPulse** (http://www.pixpulse.com/) allows users to "upload photos into your PixPulse created channel and have them show up in your other blogs and social sites."

- **Radar** (http://www.radar.net/) allows users to "upload, share and comment on yours and your friend's photos. Access through a mobile application."

- **Sharpcast** (http://www.sharpcast.com/) allows users to "show off your entire PC photo collection right on your phone. Just import your photos through Sharpcast Web Albums or the Desktop Organizer, and they automatically appear on your phone, to take with you in your pocket. And pictures you take on your phone automatically appear on your PC and the Web. You don't even have to touch a button."

- **ShoZu** (http://www.shozu.com/) allows users to "easily move photos, videos and music on and off your phone. Share them with just about every major photo hosting and blogging site."

- **Treemo** (http://www.treemo.com/) allows users to "upload photos, videos, and audio right from their phones to your treemo profile. Subscribe to your friends' channels to be updated when new content is posted."

CASHLESS WEB 2.0

- **Amazon Payments** (https://payments.amazon.com/) "provides an easy, fast, and safe means for buyers to pay you for your Marketplace and Auctions sales on Amazon.com."
- Amazon Payments Attracts Buyers: Buyers know that Amazon Payments ensures that they can pay quickly and get their goods quickly.
- Amazon Payments Is Convenient: When you're paid via Amazon Payments, there's no need for you to do anything but ship your item. Amazon.com deposits the buyer's money into your Amazon account and notifies you via e-mail. Funds in your Amazon account are deposited directly into your bank account every two weeks.
- Amazon Payments Puts You in Control: Amazon Payments helps you manage your online sales.
- **Google Checkout** (http://checkout.google.com/) provides users "a faster, safer and more convenient way to shop online."
- Stop creating multiple accounts and passwords. With Google Checkout you can quickly and easily buy from stores across the Web and track all your orders and shipping in one place.
- Shop with confidence. Our fraud protection policy covers you against unauthorized purchases made through Google Checkout, and we don't share your purchase history or full credit card number with sellers.
- Control commercial spam. You can keep your e-mail address confidential, and easily turn off unwanted e-mails from stores where you use Google Checkout.
- **KushCash** (http://www.kushcash.com/) "is a simple way to send and receive money via your mobile phone. Eliminate the usual banking hassles with our safe and easy-to-use KushCash interface. You can pay

bills and settle IOUs without having to run across town or stand in a bank line."

- **Obopay** (https://www.obopay.com/) allows users to "use any mobile phone to get, send, or spend money."

- **PayPal** (http://www.paypal.com/) "is designed from the ground up to be one of the safest ways to send money online. Unlike other financial institutions, our payments are sent without sharing financial information. In fact, PayPal never shares your financial information with or sells it to merchants. Your sensitive financial information is securely stored on our servers. When you use PayPal to pay online, you provide only your PayPal e-mail address. The merchants/retailers receive payment from PayPal without ever seeing your financial information."

- **PayPal Mobile** (https://mobile.paypal.com) allows users to "send money securely, easily, and quickly. Just reach for your mobile phone. No cash, checks, ATMs, or hassle."

- **TextPayMe** (https://textpayme.amazon.com/sdui/sdui/about) allows users "to send and receive money with your phone via text messaging."

GLOSSARY OF TERMS

The following definitions are from the author rather than a more reliable source. The intent is to provide the reader with enough information to be able to read this book without a dictionary. If you plan to appear on *Are You Smarter Than a 5th Grader*, these definitions will suffice. If you are writing your dissertation in electrical engineering, find a better source.

1G

First-generation wireless. Those bricklike analog mobile phones now available only on eBay as collector's items.

2G

Second-generation wireless. Digital mobile networks and a phone that supports text messaging.

3G

Third-generation wireless. Digital mobile networks that support wireless broadband data rates.

3.5G

Third-generation wireless plus—everything found in 3G, but with greatly increased broadband data rates that approach the speeds found in traditional DSL or cable networks. Not yet available in the United States.

4G

Fourth-generation wireless. Digital mobile networks that will offer tremendously fast wireless broadband data rates.

802.11

Those hot spots in Starbucks and most airports where you can connect to the Internet with your laptop, also known as Wi-Fi. It is the name given to local area networks by the Institute of Electrical and Electronic Engineers (IEEE—pronounced *eye triple ee* by those in the know). 802.11a, 802.11b, 802.11g, and 802.11n differ primarily in the amount of bandwidth they offer the user.

802.16

IEEE term for fixed WiMAX. Competes with DSL and cable to bridge the last mile to your home.

802.16e

IEEE term for mobile WiMAX. Competes with mobile technology, like HSDPA.

analog

Not digital. Not a series of zeros and ones. Think of a wristwatch with two hands— that is an analog device. So is a fax machine. In Negroponte terms, analog equals atoms and digital equals bits.

ARPU

Short for *average revenue per user*. It is the average amount an MNO makes on each customer.

Automated Clearinghouse (ACH)

An electronic system where a data processing center handles payments exchanged between financial institutions, primarily through telecommunications networks. ACH systems process large volumes of individual payments electronically, including salaries, consumer and corporate bill payments, interest and dividend payments, and Social Security payments.

bandwidth

A unit of measure. More bandwidth equals faster connection speeds to the Internet. You can never be too rich or have too much bandwidth.

blog

Short for *Web log*. A frequently updated online diary. An example of a Web 2.0 application.

> (In January 2006, a German politician) sent an internal e-mail to his colleagues in which he called blogs "the toilet walls of the Internet" and wanted to know, "What on earth gives every computer-owner the right to express his opinion, unasked for?" When bloggers got hold of this e-mail, they answered his question with such clarity that he quickly and publicly apologized and retreated. Inadvertently, he had put his finger on something big: that, at least in democratic societies, everybody does have the right to hold opinions, and that the urge to connect and converse with others is so basic that it might as well be added to life, liberty, and the pursuit of happiness . . . Just as everybody has an e-mail account today, everybody will have a blog in five years.[172]

Bluetooth

A short-range alternative to a cable. You've seen people with a device stuck in their ear apparently talking to themselves? That is a Bluetooth headset connected wirelessly to a mobile phone on the user's hip or very nearby.

broadband

A lot of bandwidth. For traditional wired access (DSL or cable), broadband is more than 1.5 million bits per second (Mbps). Soon 5 Mbps will be the minimal amount to be considered broadband. For wireless in the United States, .384 Mbps is considered wireless broadband.

B2B

Short for *business-to-business*.

B2C

Short for *business-to-consumer*.

C2C

Short for *consumer-to-consumer*.

Communications Assistance for Law Enforcement Act (CALEA)

A 1994 law that requires wired and wireless service providers to allow the U.S. government—with proper authorization—to listen to your phone calls.

---

172 *It's the Links, Stupid.* (2006). p. 6.

cashless

No paper or coins. Credit and debit cards are cashless, but ultimately other, more secure devices will replace cards.

cellular

The dominant mobile phone networks deployed today. Future mobile networks will use Internet protocol (IP) technology found in Internet networks rather than what we see today in cellular networks.

convergence

This term has two different meanings. The first is device convergence, the idea that a phone, television, and game console are roughly the same device in terms of what is inside them and therefore the related functions can be performed by one gadget. The iPhone is an example of device convergence. The second is service convergence, the blend of computing and telecommunications, which is often referred to now as information technology (IT). VoIP is an example of service convergence.

data

Information that has been translated into a digital form (e.g., audio, video, documents) to make it more convenient to send or store.

Defense Advanced Research Projects Agency (DARPA)

A research and development arm of the U.S. Department of Defense. DARPA invented the Internet right before Al Gore did.

digital

Not analog. Zeros and ones. Unlike analog, digital content can be compressed and stored on various media, like compact discs (CDs), digital video discs (DVDs), and MP3 players. In Negroponte terms, analog equals atoms and digital equals bits.

digital divide

The gap between those that have access to technology and those who don't.

Digital Rights Management (DRM)

This is technology, usually in the form of software, to prevent a user from sharing copyrighted content with other users.

Digital Subscriber Line (DSL)

The technology that the phone company uses to deliver broadband access to your home over existing copper wires. Unlike its predecessor, dial-up, with DSL subscribers can talk on the phone while they are accessing the Internet.

disruptive technology

The term comes from Clayton Christensen's 1997 book T*he Innovator's Dilemma*. "Disruptive technologies bring to a market a very different value proposition than had been available previously. Generally, disruptive technologies under-perform established products in mainstream markets. But they have other features that a few fringe (and generally new) consumers value" (p. xv). Usually only early adopters start playing with a disruptive technology like the telephone, computer, or mobile phone. But then prices fall and the device goes mainstream. One of the definitions of disruption is not just that it displaces someone else, but that it opens up new markets that weren't being addressed before.

> That's the way it is with disruptive technologies: almost no one comprehends what a shock to the existing order they represent until well after the rattling has begun. Do you think those guys in the cave knew what they were really onto when they clacked some rocks together and got fire?[173]

download

To receive a file from the Internet and store it on a personal device.

e-cash

Short for *electronic cash*, E-cash is a payment mechanism designed for the Internet. It is electronic money that can be passed along from person to person like cash. It is anonymous like cash, and has value immediately—it's money, not a promise to pay later.

Electronics Benefits Transfer (EBT)

A type of electronics funds transfer where public entitlement payments (e.g., welfare and food stamps) are paid to the beneficiaries through direct deposit. The beneficiaries can be given a debit card to use to spend those funds.

Fiber to the Home (FTTH)

Sometimes referred to as FTTP (Fiber to the Premise), this is a way to bring the highest levels of wired broadband right to your door (no data pipe today is as fat as a fiber optic cable) instead of gaining access to the Internet via dial-up, DSL, cable, or WiMAX.

High-Speed Downlink Packet Access (HSDPA)

A 3.5G wireless technology being used in Japan and South Korea to provide more bandwidth to subscribers.

---

173 Schonfeld, Erick, and Borzo, Jeanette. (2006, November 9). *The Next Disruptors*. Retrieved May 12, 2008, from http://money.cnn.com/magazines/business2/business2_archive/2006/10/01/8387096/index.htm.

IP

Internet Protocol. The rules that govern how data are shared over the Internet.

last mile

There is a cliché that says, "A chain is only as strong as the weakest link." The last mile is often the weakest link between your computer and the fiber optic cables that carry the fastest Internet traffic. The last mile bottleneck results from the access technology—dial-up, DSL, cable, or WiMAX—that connects your home to the fiber optic backbone having a slower than the data rate available from fiber optics. The way to fix this bottleneck is with Fiber to the Home (FTTH), also known as Fiber to the Premise (FTTP).

load balancing

Technology that allows a device that is busy to offload some of its responsibilities to another device in the network.

mobile banking

One of the financial services offered as a part of mobile commerce. Also known as *m-banking*. Refers to the availability of banking and financial services through the mobile technology, including bank and stock market transactions, mobile remittances, microfinance, and micropayments.

mobile commerce

The delivery of transaction services over mobile devices for the exchange of goods and services between consumers, merchants, and financial institutions.

m-payments

Short for *mobile payments*, m-payments are payments made using mobile devices either to directly purchase or to authorize the purchase of goods and services.

m-wallet

Short for *mobile wallet*, an m-wallet allows a mobile user to store credit or debit card information on his phone's SIM card.

Mobile Network Operator (MNO)

The more formal term for your cellular service provider, also known as cellular operator or carrier. As some service providers move to noncellular networks, this term will become more standard.

Mobile Virtual Network Operator (MVNO)

A company that acts as your mobile service provider, but does not in fact own the network you are using. MVNOs rent access from MNOs and act as sales agents building their own brand (e.g., Virgin Mobile in the United States).

MP3

Short for Moving Pictures Experts Group Audio Layer 3. MP3s allow for digital audio files to be compressed, while still maintaining close to their original sound quality.

Multimedia Messaging Service (MMS)

MMS is the video equivalent of SMS. Instead of just texting, users can share video content wirelessly with each other via their mobile devices.

multiplexed

Multiple conversations going on at the same time on the same radio channel.

Negroponte Shift

Describes the migration from devices that traditionally used wires (e.g., telephones) to wireless and the roughly simultaneous migration of devices that traditionally were wireless (e.g., television) to wire (e.g., cable television).

net neutrality

The ongoing debate over who owns the Internet. Service providers, like phone (DSL) and cable companies, want to be able to give preferential treatment—in exchange for a fee—to some content providers. Advocates of net neutrality want Internet service providers to treat all content equally.

opt-in services

Unlike e-mail spam or those annoying telemarketers, with opt-in services users choose to receive information or services from vendors. Tired of hearing about a product or service? Users can opt out.

paperless

I don't know exactly what paperless means, but I do know it doesn't mean no paper. It doesn't mean the end of books, magazines, or newspapers (yet).

peer-to-peer networks

Networks that allow users to connect directly to each other rather than going only through cell towers, switches, or Internet routers.

Personal Identification Number (PIN)

A PIN is a secret code that enables access to the data contained, for example, on a credit, debit, or smart card.

place shifting

Receiving content in a place of one's choosing (e.g., watching television on a mobile device on the subway instead of in one's living room).

podcast

A combination of *iPod* and *broadcasting*. A method of publishing audio files to the Internet for playback on mobile devices and personal computers.

Public Switched Telephone Network (PSTN)

This is the good old-fashioned telephone network, which includes the twisted copper pair of wires used in your home to make calls. You used the PSTN when accessing the Internet via a dial-up service before you switched to DSL or cable.

quadruple play

The possibility that some service providers will be able to bundle four key products together—Internet access, mobile phone service, television, and traditional phone service.

Radio Frequency Identification Devices (RFID)

RFID tags can be as small as a grain of rice and are read by RFID readers to retrieve the information stored on them. Large retailers are pressuring manufacturers to put RFID tags in all merchandise so that stores can manage their global supply chains and in-store inventory control. Passive RFID tags have no battery and must be read at very close range. Active RFID tags, like the toll tag used for driving through toll plazas in your car, do have a battery and can be read at greater range.

Really Simple Syndication (RSS)

This allows users to opt in to be notified when, for example, a blog is updated. This means the user doesn't have to keep going back to the site to see if anything new has been posted.

Secure Socket Layer (SSL)

A commonly used protocol for managing the security of message transmission on the Internet.

Search Engine Optimization (SEO)

Tricks of the trade that manipulate the results on search engines so that certain websites will appear when certain search terms are entered. Most people never go beyond the first page of their search results, so SEO is used to try to make sure your site is on that first page.

Short Message Service (SMS)

What most of the world outside the United States calls *text messaging*.

SIM

Short for subscriber identity module or subscriber information module, a SIM is a programmable card that stores all of a mobile-phone subscriber's personal information and phone settings. The card stores the phone number and personal security key necessary for the device to work. A subscriber can switch the card from phone to phone, which makes the new phone receive all calls to the subscriber's existing phone number. A SIM card also makes it possible to roam (use an MNO that you don't have an account with) around the world when the SIM card is transferred to a phone that uses the correct frequency band for that country.

smart card

Sometimes called chip cards, smart cards contain a computer chip with up to 8 KB of memory embedded in the traditional plastic credit card.

A smart card is a card that is embedded with either a microprocessor and a memory chip or only a memory chip with non-programmable logic. The microprocessor card can add, delete, and otherwise manipulate information on the card, while a memory-chip card (e.g., prepaid phone cards) can undertake only a pre-defined operation. Smart cards, unlike magnetic stripe cards, can carry all necessary functions and information on the card. Therefore, they do not require access to remote databases at the time of the transaction.[174]

smartphone

A mobile device that allows users to add applications and services. Examples would be Apple's iPhone, Motorola's Q, and Samsung's Blackjack.

Software as a Service (SAAS)

SAAS is software that resides on a server that users can access while online, as an alternative to purchasing a license to use software that resides on a user's hard drive.

---

174 Retrieved January 5, 2008, from http://java.sun.com/products/javacard/smartcards.html.

spectrum

Remember the mnemonic ROY G BIV for the colors of the rainbow (red, orange, yellow, green, blue, indigo, and violet)? These are actually frequencies of the visible segment of the electromagnetic spectrum. The rest of the frequencies of the electromagnetic spectrum are used to carry signals for everything from mobile phones to garage door openers, AM and FM radio, and broadcast television.

streaming

If you are accessing content (e.g., a YouTube video) from the Web, it is streamed to you and you must be connected to—and stay connected to—the Internet to receive it. This is different from when that content is stored on your hard drive and can be retrieved without being connected to the Internet.

third screen

Receiving advertisements on your mobile device. The first screen was your television. The second screen was your computer. Now you can be annoyed anywhere, anytime.

time shifting

Receiving content at a time of one's choosing (e.g., watching a television show that was recorded on a digital video recorder after it originally aired on broadcast television).

upload

To send a file from a personal device to the Internet.

Video on Demand (VOD)

Rather than getting content at fixed times, the user can select when to, for example, watch a movie on television. VOD is pronounced *vee oh dee* by those in the know.

Voice over Internet Protocol (VoIP)

A technology that allows one to make phone calls over the Internet rather than the PSTN. This is the kind of service provided by Skype, Vonage, and many cable companies. VoIP is pronounced *vee oh eye pea* by those in the know, and pronounced *vope* by those who don't know better.

Web 2.0

As opposed to the original World Wide Web, Web 2.0 is a set of applications that allow users to generate their own content, upload it for others to share, and interact with the contributions of others. If Web 1.0 was one-to-many, Web 2.0 is many-to-many. Web 2.0 is sometimes referred to as new media by the old media.

Fifty-seven percent of American teenagers create content for the Internet—from text to pictures, music and video. People no longer passively consume media, which usually means creating content in whatever form and on whatever scale. This does not have to mean that people write their own newspaper; it could be as simple as rating the restaurants they went to or the movie they saw, or as sophisticated as shooting a home video . . . Not everything in the blogosphere is poetry, not every audio podcast is a symphony, and not every entry on Wikipedia is 100 percent correct. But exactly the same could be said about newspapers, radio, television and the Encyclopedia Britannica.[175]

WiBro

The South Korean version of 802.16e/WiMAX.

Wi-Fi

Those hot spots in Starbucks and most airports where you can connect to the Internet with your laptop. Wi-Fi is short for *wireless fidelity*. Geeks call it 802.11 or WLAN.

wiki

User-generated and user-edited content, usually for an information-sharing purpose, such as a how-to guide or encyclopedia. The most prominent example is Wikipedia. It is based on the notion of the wisdom of crowds, which is based on the notion that there is wisdom in crowds, which has never been proven.

WiMAX

Short, fortunately, for Worldwide Interoperability for Microwave Access. One version of WiMAX allows you to connect to the Internet from your home wirelessly. It will compete with DSL and cable service providers. Geeks call this 802.16 or fixed WiMAX. The other version allows you to connect to the Internet while driving around in your car. Geeks call this 802.16e or mobile WiMAX.

wireless

No wires. Mobile phones are wireless. So are cordless phones, satellite radio, Bluetooth headsets, microwave relay stations, and RFID tags. Wireless does not necessarily mean mobile—think about the satellite television dish on your roof. Wireless does not necessarily mean cellular—think about your cordless phone in the kitchen. Wireless does not necessarily mean voice—think about text messaging. And wireless does not necessarily mean person-to-person or person-to-machine—it can be device-to-device, like RFID.

---

175 *Among the Audience.* (2006). p. 4.

WLAN

Short for *wireless local area network*, it is another term for Wi-Fi and 802.11 networks, which are what you can connect to at Starbucks, for example.

REFERENCES AND ADDITIONAL READING

THE FUTURE OF WIRELESS

A World of Connections. (2007, April 28). *The Economist* (supplement), pp. 3–4.

Among the Audience. (2006, April 22). *The Economist,* pp. 3–5.

Anders, George. (2008, January 28). Predictions of the Past. *The Wall Street Journal.* p. R3.

A Bank in Every Pocket. (2007, November 15). *The Economist.* Retrieved May 1, 2008, from http://www.economist.com/opinion/displaystory.cfm?story_id=10133998.

Bringing the Poor Online. (2008, February 22). *The Economist.* Retrieved May 1, 2008, from http://www.economist.com/daily/columns/techview/displaystory. cfm?story_id=10748746.

Bruno, Lee. (2002, February). Building the Real Information Superhighway. *Red Herring,* pp. 66–67.

Bugging the Cloud. (2008, March 8). *The Economist,* pp. 28–30.

Bures, Frank. (2007, September). Access Denied. *Wired,* pp. 60–61.

Burns, Enid. (2008, April 2). U.S. Mobile Phone Users Talking, Texting More. *ClickZ.* Retrieved May 1, 2008, from http://www.clickz.com/showPage.html?page=3628985.

Cell Phone Usage Continues to Increase in the USA. (2008, April 4). *Cellular News.* Retrieved May 1, 2008, from http://www.cellular-news.com/story/30323. php?source=newsletter.

Compose Yourself. (2006, April 22). *The Economist,* pp. 7–9.

Consumers and Convergence. (2007). KPMG. Retrieved May 1, 2008, from
    http://www.kpmginsiders.com/pdf/070725_ConsmrConvrg_POST.pdf.
De Waele, Rudy. (2006, December 11). *Understanding Mobile 2.0.* Retrieved May 1, 2008,
    from http://www.readwriteweb.com/archives/understanding_mobile_2.php.
Delivering the Bits. (2008, April 18). *The Economist.* Retrieved May 1, 2008,
    from http://www.economist.com/research/articlesBySubject/displayStory.
    cfm?story_id=11074015&subjectID=348963&fsrc=nwl.
Dredge, Stewart. (2007, October 10). Five ways RFID is being used in mobile phones. *Tech
    Digest.* Retrieved May 1, 2008, from http://techdigest.tv/2007/10/
    koreajapan_week_14.html.
Drivers Expect Telematics to Promote Security. (2002). Retrieved April 1, 2002, from
    http://www.nikkeibp.asiabiztech.com/wcs/leaf?CID=onair/asabt/resch/177664.
Dropped Call. (2008, March 7). *The Economist.* Retrieved May 1, 2008, from http://www
    .economist.com/daily/columns/techview/displaystory.cfm?story_id=10830025.
Edwards, Cliff. (2008, April 14). So Maybe Apple Was Onto Something. *Business Week*, pp.
    51–52.
Edwards, Cliff, and Moon, Ihlwan. Upward Mobility. (2006, December 4). *Business Week*,
    pp. 69–70.
Evans, Mike. (2008, March 28). New Pantech Concept Phones for 2010. Mobile Mentalism.
    Retrieved May 1, 2008, from http://mobilementalism.com/2008/03/28/
    new-pantech-concept-phones-for-2010/.
Ferguson, Robert. (2006, December). Welcome. *Charged*, p. 1.
Foust, Dean. (2007, December 3). Mobile Phones, Immobile Cars. *Business Week*, p. 68.
Frenzel, Louis. (2007, September 13). 3G Takes Charge, but 4G Looms Large. *Electronic
    Design*, pp. 48–54.
From Major to Minor. (2008, January 10). *The Economist.* Retrieved May 1, 2008, from
    http://www.economist.com/business/displaystory.cfm?story_id=10498664.
Gadgets at Work: The Blurring Boundary between Consumer and Corporate Technologies.
    (2008, April 16). *Knowledge@Wharton.* Retrieved May 1, 2008, from http://knowledge
    .wharton.upenn.edu/article.cfm?articleid=1937.
Gardner, W. David. (2007, July 3). South Korea First to Dial Motorola's iPhone challenger.
    EETimes Asia. Retrieved May 1, 2008, from http://www.eetasia.com/ART_
    8800470714_499488_NT_518410f1.HTM.
Garfinkel, Simon. (2008, March/April). Android Calling. *Technology Review*,
    pp. 88–90.
Homo Mobilis. (2008, April 10). *The Economist.* Retrieved May 1, 2008, from
    http://www.economist.com/surveys/displaystory.cfm?story_id=10950487.
The Impact of Digitization. (2007). KPMG. Retrieved May 1, 2008, from
    http://www.kpmg.ca/en/industries/ice/documents/TheImpactOfDigitalization.pdf.
International Telecommunications Union statistics. (2006). Retrieved May 1, 2008, from
    http://www.itu.int/ITU-D/ict/statistics/ict/index.html.
It's the Links, Stupid. (2006, April 22). *The Economist,* pp. 5–6.
Jaokar, Amit, and Fish, Tony. (2006). *Mobile Web 2.0.* UK: Futuretext.
Karlgaard, Rich. (2007, November 27). Our Challenge Is Change, Not Globalization. *Forbes.*
    Retrieved May 1, 2008, from http://www.mywire.com/pubs/Forbes/
    2006/11/27/1966852.
Labour Movement. (2008, April 10). *The Economist.* Retrieved May 1, 2008, from
    http://www.economist.com/research/articlesBySubject/displayStory.cfm
    ?story_id=10950378&subjectID=348963&fsrc=nwl.

Leiner, Barry M., Cerf, Vinton G., et al. A Brief History of the Internet. Retrieved May 1, 2008, from http://www.isoc.org/Internet/history/brief.shtml.

Location, Location, Location. (2008, April 10). *The Economist.* Retrieved May 1, 2008, from http://www.economist.com/research/articlesBySubject/displayStory.cfm ?story_id=10950439&subjectID=894408&fsrc=nwl.

Locke, Christopher, and Lewis, Jennifer. (2002, February). A Different Kind of Mobile Computing. *Red Herring,* p. 71.

Lowry, Tom. (2007, December 17). The FCC's Broadband Bobble. *Business Week,* p. 76.

Mann, Charles. (2004, July/August). A Remote Control for Your Life. *Technology Review,* pp. 42–49.

Marconi's Brainwave. (2007, April 28). *The Economist* (supplement), pp. 4–6.

Mobile Access to Data and Information. (2008, March). Pew Internet and American Life Project. Retrieved May 1, 2008, from http://www.pewinternet.org/PPF/r/244/report_ display.asp.

Mobile Phones Could Soon Rival the PC as World's Dominant Internet Platform. (2006, April 18). Retrieved May 1, 2008, from http://www.ipsos-na.com/news/pressrelease.cfm ?id=3049.

Nadarajan, Bem. (2008, March 2). The Face That Launched 5.5 Million Cell Phone Alerts. *The Straits Times.* Retrieved May 1, 2008 from http://www.straitstimes.com/Free/Story/ STIStory_212374.html.

Negroponte, Nicholas. (1995). *Being Digital.* New York: Random House.

Nomads at Last. (2008, April 12). *The Economist.* Retrieved May 6, 2008, from http://www .economist.com/research/articlesBySubject/displayStory.cfm?story_id=10950394.

Our Nomadic Future. (2008, April 10). *The Economist.* Retrieved May 1, 2008, from http://www.economist.com/research/articlesBySubject/displayStory.cfm ?story_id=11016402&subjectID=894408&fsrc=nwl.

Overcoming Hang-Ups. (2007, April 28). *The Economist* (supplement), pp. 6–9.

The Phone of the Future. (2006, December 2). *The Economist* (supplement), pp. 18–20.

Playing Tag. (2007, December 8). *The Economist* (supplement), p. 10.

Put on Hold. (2008, March 5). *Knowledge@Wharton.* Retrieved March 15, 2008, from http://knowledge.wharton.upenn.edu/article.cfm?articleid=1910.

Reiss, Spencer. (2007, December 20). Q&A: Author Nicholas Carr on the Terrifying Future of Computing. *Wired.* Retrieved May 1, 2008, from http://www.wired.com/ techbiz/people/magazine/16-01/st_qa.

Rheingold, Howard. (2003). *Smart Mobs.* Cambridge, MA: Perseus Publishing.

Scanlon, Jessie. (2007, September). Cross Pollinators. *InDesign,* p. 8.

Shipley, David, and Schwalbe, Will. (2007). *Send.* New York: Alfred A. Knopf.

Smith, Lee. (2007, September). NASDAQ CEO Executive Insights. *Business 2.0.*

Tanner, John C. (2007, February). Children of the Evolution. *Charged,* pp. 48–54.

Thinking About Tomorrow. (2008, January 28). *The Wall Street Journal.* pp. R1–4.

What Sort of Revolution? (2006, April 22). *The Economist,* pp. 15–16.

When Everything Connects. (2007, April 28). *The Economist,* p. 11.

Wildstrom, Stephen. (2007, December 17). Microsoft's Nifty Digital Shoebox. *Business Week,* p. 75.

Wildstrom, Stephen. (2008, February 25). PC Power in Your Handheld. *Business Week.* Retrieved May 1, 2008, from http://www.businessweek.com/magazine/content/08_08/ b4072000354662.htm.

Your Television Is Ringing. (2006, October 14). *The Economist* (supplement), pp. 1–20.

Vail, Michael. (2007, December 20). *IBM Reveals Five Innovations That Will Change Our Lives Over the Next Five Years*. Retrieved May 1, 2008, from http://www.thought-criminal.org/article/node/1103.

THE FUTURE OF PAPERLESS

Alfred, Randy. (2008, March 11). The Emperor's Court Is No Longer a Paperless Office. *Wired*. Retrieved May 1, 2008, from http://stag2.wired.com/science/discoveries/news/2008/03/dayintech_0311.

Anthony, Joseph. *Six Tips for a Paperless Office*. Retrieved May 1, 2008, from http://www.microsoft.com/smallbusiness/resources/technology/communications/6_tips_for_a_paperless_office.mspx.

As Businesses Go Paperless, Owners Face New Threats, Decisions. (2008, March 27). *MarketWatch*. Retrieved May 1, 2008, from http://www.marketwatch.com/news/story/businesses-go-paperless-owners-face/story.aspx?guid=%7BAAABA9FA-4925-4430-8DF4-1FDDB5189D24%7D.

Bradley, Matt. (2005, December 12). Whatever Happened to the Paperless Office? *The Christian Science Monitor*. Retrieved May 1, 2008, from http://www.csmonitor.com/2005/1212/p13s01-wmgn.html.

Costs and Benefits of Health Information Technology. (2006, April). The Agency for Healthcare Research and Policy. Retrieved May 1, 2008, from http://www.ahrq.gov/downloads/pub/evidence/pdf/hitsyscosts/hitsys.pdf.

Death by E-mail. (2007, October 30). Retrieved May 1, 2008, from http://gdgrifflaw.typepad.com/home_office_lawyer/paperless_office/index.html.

Do I Really Need a Paper Shredder to Protect Myself from Identity Theft? (2006, December). *Wired*, Retrieved May 1, 2008, from http://www.wired.com/wired/archive/14.12/start.html?pg=7.

Epstein, Jason. (2008, March/April). What's Wrong with the Kindle. *Technology Review*, pp. 12–13.

Expanding E-Government. (2006, December). Retrieved May 1, 2008, from http://www.whitehouse.gov/omb/egov/g-7-expanding.html.

Fairfield, Hannah. (2008, February 10). Pushing Paper Out the Door. *The New York Times*. Retrieved May 1, 2008, from http://www.nytimes.com/2008/02/10/business/10metrics.html.

Federal Reserve Studies Confirm Electronic Payments Exceed Check Payments for the First Time. (2004, December 6). Federal Reserve Financial Services Policy Committee press release. Retrieved May 1, 2008, from http://www.Federalreserve.gov/boarddocs/press/Other/2004/20041206/default.html.

Flat Prospects. (2007, March 15). *The Economist*. Retrieved May 1, 2008, from http://www.economist.com/business/displaystory.cfm?story_id=8856093.

From Clipboards to Keyboards. (2007, May 19). *The Economist*, p. 68. Retrieved May 1, 2008, from http://www.economist.com/business/displaystory.cfm?story_id=9196289.

Gardner, W. David. (2007, March 13). MIT to Put Its Entire Curriculum Online Free of Charge. *Information Week*. Retrieved May 1, 2008, from http://www.informationweek.com/news/showArticle.jhtml?articleID=198000568.

Gates, Bill. (2006, April 7). How I Work. *Fortune Magazine*. Retrieved May 1, 2008, from http://money.cnn.com/2006/03/30/news/newsmakers/gates_howiwork_fortune/.

Goudreau, Jenna. (2007, September 3). Making Digital Books into Page Turners. *Business Week*, p. 75.

Grossman, Lev. (2007, April 30). Reading Gets Wired. *Time Magazine*. Retrieved May 1, 2008, from http://www.time.com/time/magazine/article/0,9171,1612700,00.html.

In Pursuit of Paperless Personal Finance. (2007, August 27). Retrieved May 1, 2008, from http://www.getrichslowly.org/blog/2007/08/27/in-pursuit-of-paperless-personal-finance/.

Japan's Latest Mobile Craze: Novels Delivered to Your Handset. (2007, May 24). *The Economist*. Retrieved May 1, 2008, from http://www.economist.com/business/displaystory.cfm?story_id=E1_JNPTRGD.

Keen, Andrew. (2007, Summer). Against Open Culture. *AlwaysOn*, pp. 3–4.

Kher, Unmesh. (2007, March 30). Chasing Paper from Medicine. *Time Magazine*. Retrieved May 1, 2008, from http://www.time.com/time/magazine/article/0,9171,1604858,00.html.

Kiley, David. (2007, December 4). Amazon Can Empty Bookstore Shelves. *Business Week*. Retrieved May 1, 2008, from http://www.businessweek.com/technology/content/dec2007/tc2007123_374610.htm?campaign_id=techn_Dec4&link_position=link37.

Koppel, Ross, et. al. (2005, March 9). Role of Computerized Physician Order Entry Systems in Facilitating Medication Errors. *Journal of American Medicine*. Retrieved May 1, 2008, from http://jama.ama-assn.org/cgi/content/abstract/293/10/1197.

Lazaroff, Leon. (2007, February 23). Software Lets Readers Download Newspapers. *The Orlando Sentinel*. Retrieved May 26, 2007, from http://www.orlandosentinel.com/technology/orl-hearst2307feb23,0,5444086.story.

LeClaire, Jennifer. Will Amazon Kindle an E-Book Fire? *Newsfactor.com*. Retrieved May 1, 2008, from http://www.newsfactor.com/story.xhtml?story_id=11100CLJ4PV0&nl=2.

Levy, Stephen. (2007, November 26). The Future of Reading. *Newsweek*, pp. 57–64.

Levy, Stephen. (2007, November 26). Can It Kindle the Imagination? *Newsweek*, p. 64.

The Most Deadly Weapon in My Technology Quiver. (2006, October 9). Retrieved May 1, 2008, from http://greatestamericanlawyer.typepad.com/greatest_american_lawyer/paperless_law_office/index.html.

Offner, Jim. (2008, February 14). Digital Downloading at Heart of Borders' Splashy New Concept Stores. *Business Week*. Retrieved February 25, 2008, from http://www.technewsworld.com/rsstory/61679.html.

Openshaw, Jennifer. (2007, July 4). The 15-Minute Tip. *MarketWatch*. Retrieved May 1, 2008, from http://www.marketwatch.com/news/story/15-minute-tip-should-you-put/story.aspx?guid=%7B7449896B-3DEF-4983-BB7F-E683412F9D89%7D.

Paperless Budget Will Save Cash, Trees (2008, January 10). Orlando Sentinel, p. A7. Retrieved January 10, 2008, from http://www.orlandosentinel.com/technology/orl-budget1008jan10,0,6562792.story.

The Paperless Library. (2005, September 22). *The Economist*. Retrieved May 1, 2008, from http://www.economist.com/science/displaystory.cfm?story_id=E1_QQNPGQG.

Perkowski, Mateusz. (2006, September 6). Practically Paperless Lawyer. *The Forest Grove News-Times*. Retrieved May 1, 2008, from http://www.forestgrovenewstimes.com/sustainable/story.php?story_id=115756852070690900.

Robinson, Rhonda. (2007, August 22). Winning the Race to a Paperless Office. *Fulton County Daily Report*. Retrieved May 1, 2008, from http://www.law.com/jsp/legaltechnology/pubArticleLT.jsp?id=1187686937571.

Screen Savers. (2007, May 24). *The Economist*. Retrieved May 1, 2008, from http://www.economist.com/business/displaystory.cfm?story_id=9231860.

Segan, Sascha. (2007, March 14). Death to the Fax Machine. *PC Magazine*. Retrieved May 1, 2008, from http://www.pcmag.com/article2/0,1895,2102956,00.asp.

Sellen, Abigail J., and Harper, Richard. (2001). *The Myth of the Paperless Office*. Retrieved May 1, 2008, from http://mitpress.mit.edu/catalog/item/default.asp?ttype=2&tid=8501.

U.S. Plans Social Security Debit Cards. (2008, January 4). Retrieved May 1, 2008, from http://www.upi.com/NewsTrack/Top_News/2008/01/04/us_plans_social_security_debit_cards/7716/.

The Use of Checks and Other Non-Cash Instruments in the United States. (2002, August). *Federal Reserve Bulletin*. Retrieved May 1, 2008, from http://www.federalreserve.gov/pubs/bulletin/2002/0802_2nd.pdf.

Weston, Liz Pulliam. *Go Paperless for Safer Banking*. Retrieved May 1, 2008, from http://articles.moneycentral.msn.com/Banking/BetterBanking/GoPaperLessForSaferBanking.aspx.

Wildstrom, Stephen. (2007, December 3). Tech & You. *Business Week*, p. 74.

Woyke, Elizabeth. (2007, April 9). Wanted: A Clutter Cutter. *Business Week*. Retrieved May 1, 2008, from http://www.businessweek.com/magazine/content/07_15/c4029011.htm.

THE FUTURE OF CASHLESS

A Cash Call. (2007, February 17). *The Economist*, pp. 71–73.

Amin, Shaker. (2007, May 14). *M-banking—To Bank the Unbanked*. Frost & Sullivan. Retrieved May 1, 2008, from http://www.frost.com/prod/servlet/market-insight-top.pag?docid=98655381.

Ashcraft, Brian. (2008, March 28). Japanese Schoolgirl Watch: Tobacco Vending Machines Block Underage Smokers. *Wired*. Retrieved May 1, 2008, from http://www.wired.com/culture/lifestyle/magazine/16-04/st_jsgw.

Beard, Chris. *The Cashless Society Is Here*. Retrieved May 1, 2008, from http://www.geocities.com/heartland/pointe/4171/profeticword.html.

Bhengu, Xolile. (2007, November 22). 13 Million Can Cash In on Telephone Banking. *The Times*. Retrieved May 1, 2008, from http://www.thetimes.co.za/News/Article.aspx?id=626870.

Bruene, Jim. (2007, March 26). *Mobile Payments Metrics: NTT DoCoMo*. Retrieved May 1, 2008, from http://www.netbanker.com/2007/03/mobile_payments_metrics_ntt_docomo.html.

Cannon, Ellen. (2007, November 12). *2007 Gift Card Study*. Retrieved May 1, 2008, from http://www.bankrate.com/brm/news/cc/20071112_gift_card_study_analysis_a1.asp.

Cashiered. (2007, April 7). *The Economist*, p. 73.

Cashless Society Gets Mixed Reviews. CNN.com/Technology. Retrieved May 1, 2008, from http://www.cnn.com/2003/TECH/ptech/02/08/cash.smart.ap/.

Convergence Takes Hold. (2006). KPMG. Retrieved May 1, 2008, from http://www.kpmg.com.au/Portals/0/ConvergenceTakesHold-GlobalICEMarkets.pdf.

Crying Voice. Retrieved December 9, 2006, from http://www.cryingvoice.com/Endtimes/Mark12.html.

Der Hovanesian, Mara (2007, December 17). Check That Check. *Business Week*, p. 18.

Downey, Catherine M. (1996, Winter). The High Price of a Cashless Society: Exchanging Privacy Rights for Digital Cash. *The John Marshall Journal of Computer & Information Law*. Retrieved May 1, 2008, from http://www.jcil.org/journal/articles/304.html.

Dreams of a Cashless Society. (2001, May 3). *The Economist*. Retrieved May 1, 2008, from http://www.economist.com/finance/displayStory.cfm?Story_ID=613491.

Dubner, Stephen J., and Levitt, Steven D. (2007, January 7). The Gift Card Economy. *The New York Times*. Retrieved May 1, 2008, from http://www.nytimes.com/2007/01/07/magazine/07wwln_freak.t.html.

Farivar, Cyrus. (2004, August). New Ways to Pay. *Business 2.0*, p. 26. Retrieved May 1, 2008, from http://money.cnn.com/magazines/business2/business2_archive/2004/08/01/3773 81/index.htm.

Five Ways Paperless Personal Finance Saves Your Money. (2007, August 30). Retrieved May 1, 2008, from http://www.bargaineering.com/articles/5-ways-paperless-personal-finance-saves-you-money.html.

Fost, Dan. (2007, November 20). One More Thing Cell Phones Could Do: Replace Wallets. *USA Today*. Retrieved May 1, 2008, from http://www.usatoday.com/money/industries/technology/2007-11-20-paying-by-cellphone_N.htm.

Germain, Jack M. (2004, December 25). *Biometric Cell Phones on Slow Track to U.S. Market*. Retrieved May 1, 2008, from http://www.technewsworld.com/story/39134.html.

Griffith, Reynolds. (1994, March). *Cashless Society or Digital Cash?* Southwestern Society of Economists.

Levisohn, Ben. (2008, March 3). Prepaid Cards: The Cleanup. *Business Week*. Retrieved May 1, 2008, from http://www.businessweek.com/magazine/content/08_09/b4073032428110.htm.

Lev-Ram. Michal. (2006, December 1). Your Cell Phone = Your Wallet. *Business 2.0*. Retrieved May 1, 2008, from http://money.cnn.com/2006/11/30/magazines/business2/cash_cellphones.biz2/index.htm.

Lewan, Todd. (2008, January 26). *Microchips Everywhere: A Future Vision*. Retrieved February 4, 2008, from http://www.washingtonpost.com/wp-dyn/content/article/2008/01/26/AR2008012601126.html.

Malykhina, Elena. Deployments of Contactless Payment Systems Slower than Expected. (2008, January 10). *Information Week*. Retrieved May 1, 2008, from http://www.informationweek.com/news/showArticle.jhtml?articleID=205602142.

McDonald's Expands Cashless Payment Options with MasterCard PayPass. (2004, August 18). Retrieved May 1, 2008, from http://www.qsrweb.com/article.php?id=409.

Metcalf, Allan, and David K. Barnhart. *America in So Many Words*. Retrieved May 1, 2008, from http://www.houghtonmifflinbooks.com/epub/americawords.shtml.

Mobile Payments in Asia Pacific. (2007). KPMG. Retrieved May 1, 2008, from http://www.kpmginsiders.com/pdf/Mobile_payments.pdf.

Morrow, George. (1984, November). A Computerized Cashless Society. *Creative Computing*. Retrieved May 1, 2008, from http://www.atarimagazines.com/creative/v10n11/271_A_computerized_cashless_s.php.

Negroponte, Nicholas. (1995). *Being Digital*. New York: Random House.

Nokia Launches New Phone with Electronic Wallet. (2008, April 15). *Reuters*. Retrieved May 1, 2008, from http://www.reuters.com/article/technologyNews/idUSL1578787320080415.

Non-Cash Payment Trends in the United States 2003–2006. (2007, December 10). Research sponsored by the Federal Reserve System. Retrieved May 1, 2008, from http://www.frbservices.org/files/communications/pdf/research/2007_payments_study.pdf.

Overdorf, Jason. (2007, December 17). Cashless in the Hinterlands. *Newsweek*. Retrieved May 1, 2008, from http://www.newsweek.com/id/74440.

Panhandlers Beware. (2006, November 16). *The Economist*, p. 81. Retrieved May 1, 2008, from http://www.economist.com/finance/displaystory.cfm?story_id=E1_RTSPJGR.

Parmelee, Nathan. (2007, February 1). DoCoMo's Holding Strong. *The Motley Fool*. Retrieved May 1, 2008, from http://www.fool.com/investing/international/2007/02/01/docomos-holding-strong.aspx.

Privacy in Radio Frequency Identity Documents. (2006). Texas Instruments White Paper.

Rudden, Liam. (2007, September 19). *Banking on the Cashless Society*. Retrieved May 1, 2008, from http://edinburghnews.scotsman.com/liamrudden/Banking-on-the-cashless-society.3328495.jp.

Starrs, Chris. (2008, February 6). Pastor Says Cashless Society Would Fulfill Bible Prophecy. *Athens Banner-Herald*. Retrieved May 1, 2008, from http://endrtimes.blogspot.com/2008/03/cashless-society-fulfill-bible-prophecy.html.

Think Big for Cashless Banking. (2006, September 5). Retrieved May 1, 2008, from http://www.banking-business-review.com/article_feature.asp?guid=DF7415E2-3E20-4E0D-9BC6-1318991A4D70.

Vernon, Wes. *Cashless Society "Inevitable"; a Boost to Globalist Taxers?* Retrieved May 1, 2008, from http://www.newsmax.com/archives/articles/2002/6/28/181711.shtml.

The Unwired Coke Machine. (2001, September 15). *America's Network*. Retrieved May 1, 2008, from http://www.americasnetwork.com/americasnetwork/article/articleDetail.jsp?id=176.

Upward Mobility. (2006, December 4). *BusinessWeek*. Retrieved May 1, 2008, from http://www.businessweek.com/magazine/content/06_49/b4012071.htm.

Verizon Wireless Launches Mobile Banking Services. (2008, January 3). *Payments News*. Retrieved May 1, 2008, from http://www.paymentsnews.com/2008/01/verizon-wireles.html.

Warwick, David. (2004, July/August). Toward a Cashless Society. *The Futurist*. Retrieved May 1, 2008, from http://www.wfs.org/excerptja04.htm.

What Is a Smart Card? (n.d.). Retrieved May 1, 2008, from http://computer.howstuffworks.com/question332.htm.

Will 2008 Be the Year of the Smart Card? (2008, January 2). *Payments News*. Retrieved May 1, 2008, from http://www.paymentsnews.com/2008/01/will-2008-be-th.html.

Wreden, Nick. (2004). *Wireless: The Next Generation*. Retrieved December 9, 2006, from http://www.risnews.com/CSS/pages/archives/articles/art_march3.html.

U.S. Plans Social Security Debit Cards. (2008, January 4). Retrieved May 1, 2008, from http://www.upi.com/NewsTrack/Top_News/2008/01/04/us_plans_social_security_debit_cards/7716/.

Vail, Michael. (2007, December 20). *IBM Reveals Five Innovations That Will Change Our Lives Over the Next Five Years*. Retrieved May 1, 2008, from http://www.thought-criminal.org/article/node/1103.

WHAT IN THE WORLD IS GOING ON?

2007 Informatization White Paper. (2007). National Information Society Agency. South Korea. Retrieved May 1, 2008, from http://www.nia.or.kr/special_content/eng/.
A Bank in Every Pocket. (2007, November 15). *The Economist.* Retrieved May 1, 2008, from http://www.economist.com/opinion/displaystory.cfm?story_id=10133998.
Ahonen, Tomi, and O'Reilly, Jim. (2007). *Digital Korea.* London: Futuretext.
America's Emobyte Deficit. (2007, November 27). *The Economist.* Retrieved May 1, 2008, from http://www.economist.com/daily/columns/businessview/displaystory.cfm?story_id=10201521.
Amin, Shaker. (2007, May 14). *M-banking—To Bank the Unbanked.* Frost & Sullivan. Retrieved May 1, 2008, from http://www.frost.com/prod/servlet/market-insight-top.pag?docid=98655381.
Ashcraft, Brian. (2008, March 28). Japanese Schoolgirl Watch: Tobacco Vending Machines Block Underage Smokers. *Wired.* Retrieved May 1, 2008, from http://www.wired.com/culture/lifestyle/magazine/16-04/st_jsgw.
Asia Pacific Mobile Communications Outlook. (2007). Frost & Sullivan Research Paper #P076-65.
Behind the Digital Divide. (2005, March 10). *The Economist.* Retrieved May 1, 2008, from http://www.economist.com/search/displaystory.cfm?story_id=E1_PSTQDVR.
Bhengu, Xolile. (2007, November 22). 13 Million Can Cash In on Telephone Banking. *The Times.* Retrieved May 1, 2008, from http://www.thetimes.co.za/News/Article.aspx?id=626870.
Bures, Frank. (2007, September). Access Denied. *Wired,* pp. 60–61.
Calling Across the Divide. (2005, March 10). *The Economist.* Retrieved May 1, 2008, from http://www.economist.com/opinion/displaystory.cfm?story_id=3739025.
China to Relax Credit Card Market. (2007, December 19). Retrieved May 1, 2008, from http://edition.cnn.com/2007/BUSINESS/12/19/china.creditcards.ap/index.html.
Choe, Sang-Hun. (2007, December 12). To Save, Koreans Use Credit Cards. *International Herald Tribune.* Retrieved May 1, 2008, from http://www.iht.com/articles/2007/12/12/news/cards.php.
ComScore Media Metrix. (2007, January). Retrieved May 1, 2008, from http://www.websiteoptimization.com/bw/0703/.
Consumer Champion. (2007, November 8). *The Economist.* Retrieved May 1, 2008, from http://www.economist.com/specialreports/displaystory.cfm?story_id=10053224.
De Waele, Rudy. (2006, December 11). *Understanding Mobile 2.0.* Retrieved May 1, 2008, from http://www.readwriteweb.com/archives/understanding_mobile_2.php.
DeLong, J. Bradford. (2003, September). Seoul of a New Machine. *Wired,* p. 83.
Dredge, Stewart. (2007, October 10). Five ways RFID is being used in mobile phones. *Tech Digest.* Retrieved May 1, 2008, from http://techdigest.tv/2007/10/koreajapan_week_14.html.
Duerden, Charles. (2007, March/April). Convergence is Coming. *Invest Korea Journal,* pp. 7–12.
Does Not Compute. (2007, November 8). *The Economist.* Retrieved May 1, 2008, from http://www.economist.com/specialreports/displaystory.cfm?story_id=10053304.

Evans, Mike. (2008, March 28). New Pantech Concept Phones for 2010. Mobile Mentalism. Retrieved May 1, 2008, from http://mobilementalism.com/2008/03/28/new-pantech-concept-phones-for-2010/.

Gardner, W. David. (2007, July 3). South Korea First to Dial Motorola's iPhone challenger. *EETimes Asia*. Retrieved May 1, 2008, from http://www.eetasia.com/ART_8800470714_499488_NT_518410f1.HTM.

Hall, Kenji. (2008, February 25). Japan: Google's Real-Life Lab. *Business Week*. Retrieved May 1, 2008, from http://www.businessweek.com/magazine/content/08_08/b4072055361150.htm.

Hardin, Garrett (1968). *The Tragedy of the Commons*. Retrieved May 1, 2008, from http://dieoff.org/page95.htm.

Heo, Jae-Young. (2005, March/April). The Revolution on your Palm. *Invest Korea Journal*, pp. 10–17.

Ihlwan, Moon. (2007, March 12). Digital South Korea's Wireless World. *Business Week*. Retrieved May 1, 2008, from http://www.businessweek.com/globalbiz/content/mar2007/gb20070312_592167.htm.

Ihlwan, Moon. (2007, January 27). The Mobile Internet's Future Is East. *Business Week*. Retrieved May 1, 2008, from http://www.businessweek.com/technology/content/jan2007/tc20070129_681556.htm.

International Telecommunications Union statistics. (2006). Retrieved May 1, 2008, from http://www.itu.int/ITU-D/ict/statistics/ict/index.html.

Kimani, Mary. (2008, January 4). *Africa: A Bank in Every Pocket?* Retrieved May 1, 2008, from http://allafrica.com/stories/200801040770.html.

Korea: Mobile Banking Takes Off. (2004, September 27). *Business Week*. Retrieved May 1, 2008, from http://www.businessweek.com/magazine/content/04_39/b3901068.htm.

Korea's Rise as an E-Nation. (2006, May/June). *Invest Korea Journal*, pp. 8–13.

Kushida, Kenji, (2008, February 1). *Wireless Bound and Unbound: The Politics Shaping Cellular Markets in Japan and South Korea*. Berkeley Roundtable on the International Economy.

Kushida, Kenji and Oh, Seung-Youn. (2006, June 29). *Understanding South Korea and Japan's Spectacular Broadband Development Liberalization of the Telecommunications Sectors*. Berkeley Roundtable on the International Economy.

Lairson, James. (1995). *The Telecommunications Revolution in Korea*. Hong Kong: Oxford University Press.

Lakshman, Nandini. (2007, October 29). A Wireless Revolution in India. *Newsweek*. Retrieved May 1, 2008, from http://www.businessweek.com/technology/content/oct2007/tc20071026_981629.htm.

Lev-Ram, Michal. (2007, August 24). Samsung's Identity Crisis. *Business 2.0*. Retrieved May 1, 2008, from http://money.cnn.com/2007/08/06/technology/samsung.biz2/index.htm.

Man's Best Friend. (2005, April 2). *The Economist* (supplement), pp. 7–10.

Margolis, Mac. (2008, February 23). Poor Countries Yield Big Profits. *Newsweek*. Retrieved May 1, 2008, from http://www.newsweek.com/id/114685.

McLaren, Warren, (2007, January 10). Motorola's Bike Charger for Mobile Phones. Retrieved May 1, 2008, from http://www.treehugger.com/files/2007/01/motorolas_bike.php.

Ministry of Information and Communication (http://eng.mic.go.kr/eng/index.jsp).

Mobile Payments in Asia Pacific. (2007). KPMG. Retrieved May 1, 2008, from http://www.kpmginsiders.com/pdf/Mobile_payments.pdf.

Of Internet Cafes and Power Cuts. (2008, February 7). *The Economist*. Retrieved May 1, 2008, from http://www.economist.com/research/articlesBySubject/PrinterFriendly.cfm?story_id=10640716.

On the Frontier of Finance. (2007, November 15). *The Economist*. Retrieved May 1, 2008, from http://www.economist.com/opinion/displaystory.cfm?story_id=10146637.

Overdorf, Jason. (2007, December 17). Cashless in the Hinterlands. *Newsweek*. Retrieved May 1, 2008, from http://www.newsweek.com/id/74440.

Overcoming Hang-Ups. (2007, April 28). *The Economist* (supplement), pp. 6–9.

Point Topic and WebsiteOptimization.com. (2007, January). Retrieved May 1, 2008, from http://www.websiteoptimization.com/bw/0704/.

Screen Test. (2007, September 6). *The Economist*. Retrieved May 1, 2008, from http://www.economist.com/displayStory.cfm?story_id=9767747.

Switch Is On: Korea. (1999). United Nations Development Program, Sharing Innovative Experiences series, p. 1. Retrieved May 1, 2008, from http://tcdc.undp.org/sie/experiences/vol1/Switch%20is%20on.pdf.

Siwicki, Bill. (2008, March). E-commerce on the Move. *Internet Retailer*. Retrieved May 1, 2008, from http://www.internetretailer.com/article.asp?id=25537.

Tabuchi, Hiroko. (2007, November 4). PCs Being Pushed Aside in Japan. Retrieved January 10, 2008, from http://biz.yahoo.com/ap/071104/bye_bye_pcs.html.

Taylor, Chris. (2006, June 14). The Future Is in South Korea. *Business 2.0*. Retrieved May 1, 2008, from http://money.cnn.com/2006/06/08/technology/business2_futureboy0608/index.htm.

Technology in 2008. (2007, December 23). *The Economist*, Retrieved May 1, 2008, from http://www.economist.com/daily/columns/techview/displaystory.cfm?story_id=10410912.

Telecoms Korea (http://www.telecomskorea.com/).

Teves, Oliver. (2007, September 30). *Cell Phones Double as Electronic Wallets*. Retrieved May 1, 2008, from http://www.usatoday.com/tech/products/2007-09-30-3242713801_x.htm.

The Real Digital Divide. (2005, March 10). *The Economist*, Retrieved May 1, 2008, from http://www.economist.com/opinion/displaystory.cfm?story_id=3742817.

The Third Screen. (2007, July 21). *The Economist*, p. 65.

Vail, Michael. (2007, September 17). *McDonald's Uses RFID for M-Commerce . . . or Is It McCommerce?* Retrieved May 1, 2008, from http://www.thought-criminal.org/2007/09/17/mcdonalds-uses-rfid-for-m-commerce-or-is-it-mccommerce.

Wong, May. (2008, January 4). *Intel Quits One Laptop Per Child Program*. Retrieved May 1, 2008, from http://news.wired.com/dynamic/stories/I/INTEL_ONE_LAPTOP_PER_CHILD?SITE=WIRE&SECTION=HOME&TEMPLATE=DEFAULT&CTIME=2008-01-04-00-59-46.